THEY EXIST

A Review of Key Literature on Extraterrestrial Existence

by

Piers Morris

**Grosvenor House
Publishing Limited**

First Edition, April 2015
Revised Edition, November 2015

The right of Piers Morris to be identified as the author of this
work has been asserted in accordance with Section 78
of the Copyright, Designs and Patents Act 1988

The book cover picture is copyright to Piers Morris
Cover design by Kevin Yeo
Website: www.kevinyeo.com
The book cover design is entitled 'The Real World'. It depicts a
particle collision, and the study of Quantum Physics where the
real world contradicts our understanding of the perceived world.

This book is published by
Grosvenor House Publishing Ltd
28-30 High Street, Guildford, Surrey, GU1 3EL.
www.grosvenorhousepublishing.co.uk

A CIP record for this book
is available from the British Library

ISBN 978-1-78148-914-7
Website: www.piersmorris.com

About the Author

The author is a well-travelled lawyer and writer. He is British, and has worked as a solicitor for many years. He continues to work as a lawyer on international projects worldwide.

He has previously written *Risk in Business*, a book on business risk and how our conduct influences success or failure, and *Brief Notes on Existence*, which was a summary of key literature in the field of mind-body spirit.

His latest book, *They Exist,* is a review of key literature covering extraterrestrial existence, and the impact it has on our lives today.

Contents

PART 1

An Introduction to The Existence of Extraterrestrial Life and its Effect on Earth

Extraterrestrial Life on Earth

The Dalai Lama made an important statement in 1992. He stated that "naturally most of us would like to die a peaceful death, but it is also clear that we cannot hope to die peacefully if our lives have been full of violence, or if our minds have mostly been agitated by emotions like anger, attachment, fear. So if we wish to die well, we must learn to live well: Hoping for a peaceful death, we must cultivate peace in our mind, and in our way of life".

As the subject of extraterrestrial life on Earth is discussed, the importance of this statement will become increasingly obvious. Extraterrestrial existence has had an impact on our lives, even if this does not seem immediately apparent. We must pay attention to how we live our lives today, and what we allow to affect our lives today, as the world we live in is changing rapidly, often without our knowledge or acceptance.

The Dalai Lama's reference above to "lives of violence", agitation, anger, and so on brings us to the starting point for the revelations about extraterrestrial life on Earth. This is William Bramley's little-known book called *The Gods of Eden*, published in 1989, long before the surge of public interest in all things extraterrestrial. It is almost a cult word-of-mouth book, because it depicts a reality on Earth which the rulers and powers-that-be are not keen to reveal. That revelation is that violence on Earth is closely tied to the existence of the Unidentified Flying Object (UFO), and the fact that human lives are inextricably linked to, and influenced by, extraterrestrials.

Bramley says the revelation in his book concerning the role of these strange flying craft didn't start out that way. He started out by trying to understand why humans are so violent, and where all this violence comes from. But, like the many other researchers that are referenced and reviewed throughout the course of this book, he reached the one and only rational conclusion: *they exist.* Extraterrestrial existence is real.

In William Bramley's own words, the basis for his book is the seemingly outrageous suggestion that: "Human beings appear to be a slave race languishing on an isolated planet in a small galaxy. As such, the human race was once a source of labour for extraterrestrial civilisation, and still remains a possession today. To keep control over its possession and to maintain Earth as something of a prison, that other civilisation has bred never-ending conflict between human beings, has promoted human spiritual decay, and has erected on Earth conditions of unremitting physical hardship. This situation has existed for thousands of years and it continues today."

This concept is not solely the thought process of William Bramley, but has been promoted by writers and lecturers both before and after Bramley's book in one form or another. Many of the works of these writers will be referenced in this book. One of them, Timothy Good, in his outstanding work *Need to Know*, has stated that he has long subscribed to the hypothesis that *homo sapiens* is a genetically modified species, deriving from the colonization of Earth by an alien race millennia ago.

Bramley references as his own starting point a book written a century ago by Charles Fort called *The Book of the Damned*, in which Fort says that Earth is now a possession or the property of 'something'. Fort says that he accepts that "in the past, before proprietorship was established, inhabitants of a host of other worlds have dropped here, hopped here, wafted, flown, been pulled here, pushed, come singly, in enormous numbers; have visited occasionally, have visited periodically for hunting, trading, replenishing harems, mining; have been unable to stay here, have established colonies here, have been lost here; far-advanced peoples, or things, and primitive peoples or whatever they were: white ones, black ones, yellow ones..." Fort's book catalogues vast numbers of unexplained phenomena and he asks whether there might be "vast living things" in the sky that we rarely if ever see.

Bramley refers to the owners of Earth as the "Custodians", but they have been referred to by many names, including Gods, Controllers, the Elite, the Annunaki, the Nephilim, and the Illuminati. It seems there was a time when humans did not exist on Earth, at least not in today's form of 'human', and the only occupants of Earth were the owners of Earth (which in this book we shall call "the Controllers"). This was the basis of Zecharia Sitchin's classic book *The Twelfth Planet*, the substance of which derived from the translation of cuneiform clay tablets on which information about their civilisation was written. Some disagree with Sitchin's translations, but what cannot be questioned is that these now departed/lost peoples had the good sense to write on a format that would survive the test of time. Today's civilisation will not be recorded for future civilisations, as microchips and paper will not survive.

PART 2

Creation

The Reptilian Origins of Mankind

An extremely instructive source for understanding about our human origins is *Flying Serpents and Dragons* by R.A.Boulay, subtitled *The Story of Mankind's Reptilian Past*. When the first edition of his book came out in 1990, he mentioned that the idea of our ancestors being reptiles seemed to be complete fantasy. However, by the end of the nineties when the revised edition came out, thought processes had shifted sufficiently to have softened the impact of the original work, and the subject is now debated world-wide. Many now suggest openly that our ancestors may indeed have been reptilian astronauts.

Boulay's sources are an array of pseudepigraphical works that supplemented what was written in The Old Testament. Boulay states that it is now accepted fact that The Old Testament, especially the Book of Genesis, is an abbreviated version of myriads of documents ('pseudepigraphical' and 'apocryphal' documents) available to the early chroniclers who assembled the scriptures. These documents expand on the stories of the Old Testament, a rich Jewish oral tradition that is summarized under the name of "Haggadah", and other religious documents that are not written in Hebrew, but are written, in particular, in Latin, Ethiopic, Slavic and Coptic.

Other sources Boulay relies on are the large numbers of Sumerian documents brought to global prominence by Zecharia Sitchin, the King List referenced by various Sumerian scholars such as Samuel Noah Kramer and Thorkild Jacobsen, and the Gnostic literature which was eliminated from the Bible as it undermined the precepts of the male-dominated early church and stated there was no need for an intermediary between man and God. In 1945, extensive Gnostic treatises were found in earthenware jars in Egypt at a small town called Nag Hammadi.

The Sumerian documents refer to the arrival of a group of 'gods' called the Anunna (the Sumerian name) or Annunaki (the later Akkadain or Semitic name) who established a civilisation in Mesopotamia long before the Deluge. The ancient Sumerian kings traced their lineage back some 240,000 years to when they first arrived on Planet Earth. Their arrival and subsequent colonization of Earth is described in the King List, which originated in the third Millennia BC and which detailed the various dynasties that ruled in Mesopotamia over a period of 241,200 years before the rule was ended by the Flood.

The arrival of the Sumerians provided the written evidence for the sudden appearance of modern man. Boulay states that modern man seems to have been created and developed in two stages. Firstly, as a slave worker when he was mostly reptilian in appearance and nature, then secondly, when his mammal nature was increased so he lost his scaly skin and hide and could reproduce himself. He developed soft skin and hair and required clothing for comfort and protection. This stage has been described as the "Fall of Man" in

Biblical terms. The Haggadah describes the serpent-gods who were tall, at least 8 to 10 feet, and walked on two legs. They had a tail like a reptile, and a tough hide somewhat like a lizard but with a large amount of horny or scaly skin. Their hide was generally lustrous and smooth, and varied in colour from green to grey. They went around naked but wore clothing such as cloaks as a sign of rank or godship.

Boulay reports that they needed moisture and warmth, which was most likely why they founded civilisations at the mouth of great river systems: the Nile delta, the Indus river valley and the Tigris-Euphrates system. In these areas, our antediluvian Earth climate would have been to their liking as it approximated conditions on their home planet, with high amounts of carbon dioxide and moisture due to the cloud canopy. This land is described in Sumerian as "E-Din", which translates as "The Land of the Gods or Righteous Ones". As the climate changed over time and dried out, this would have brought discomfort to them. After the Deluge they went underground or left.

What is apparent from Boulay's work is that in the West we have decided to depict 'God' to be in our image, rather than any other. In this way, we have hidden the true identity of our creators. Early Chinese and Indian literature, for instance, does not have any difficulty in stating that humans were formed from dragons and serpents. Nor does Central American and African mythology.

According to Boulay's work, when the hybrid creature used by the Annunaki rulers as slave labour was created, it would have looked reptilian. The Book of Genesis in the Bible is quite specific about this, stating "then God

said 'I will make man in my image, in my likeness'".
Later, as man intermarried with his species, the reptilian
strain deteriorated and he became more mammal-like
and less and less reptilian in appearance. The mammal
genes dominated the reptilian ones and man became
more "human". This change in appearance over time
was noted in Genesis as the source of original sin and
the fall from grace, because of the loss of the reptilian
features and the inability to remain naked. The Haggadah
states that Adam and Eve "had been overlaid with a
horny skin" and Adam was "as bright as daylight and
covered his body with a luminous garment", meaning
the scaly and shiny skin. The change of appearance
caused the human to sweat, unlike a reptile, and this
is described in Genesis where Adam is told that his
punishment for "eating the forbidden fruit" was that
"by the sweat of your face shall you earn bread".

Humans were repulsed by the reptilian form. Boulay
quotes an early Gnostic document which states that,
because of their eating of the forbidden fruit, "they
saw their makers, they loathed them since they were
beastly forms". Revelations such as this in Gnostic
literature was a major cause for the persecution of
Gnostics by the fathers of the early church.

It appears that the distaste of the reptilians and their
creations for each other was mutual. According to
Boulay research, the reptilian rulers considered man
to be an inferior animal. This is made apparent from
a tract in the Hebrew Book of Enoch, where mankind
is described as "born of woman, blemished, unclean,
defiled by blood and impure flux, men who sweat putrid
drops". The Old Testament regularly notes the disgust of
the "angels" towards their sweaty mammalian cousins,

using the phraseology of "weakness of the flesh". Moreover, the mammalian form reduced lifespan as the saurian gene became more and more diluted, again referred to in Biblical terms as the "weakness of the flesh". Genesis interestingly reports an incident after the Deluge where Noah, drunk from too much wine, collapses in a drunken stupor in his tent. His son Ham finds him naked and Noah loses all sense of reason when he discovers that he has been seen naked, and puts a curse on Ham and his brother Canaan. Why this irrational behaviour? Perhaps because Noah's scaly skin was revealed?

In The Old Testament, the overriding need for privacy of the Hebrew God Yahweh (Jehovah) is made clear. On Mount Sinai, when Moses asked to see the face of his god, he was told "you cannot see my face, for man may not see me and live". Was this because the appearance of the 'god' would be repulsive to a human, and so must not be revealed?

Boulay concludes by stating that it is a most superb irony that a race of intelligent beings may exist in our neighbourhood of space who are reptilian and repulsive by our standards, and yet have founded our human civilisation. He goes on to state that a race that could traverse space would certainly have achieved genetic engineering and the ability to regenerate themselves and thereby achieve long and extended life.

Intelligent Design

As was mentioned by RA Boulay in the previous chapter, the work of Zecharia Sitchin, in a series of books starting with *The Twelfth Planet* back in 1976, referenced the creation of a new creature which was animated and able to perform arduous tasks on behalf of its masters.

One of the things that Sitchin referenced from his translation work was the fact that the Annunaki worked on genetically enhancing a native being of Earth known as the *homo erectus*. In his book *Everything You Know is Wrong*, Lloyd Pye maintains that the Annunaki, in fact, worked on a slightly more evolved version of the hominid known as the Neanderthal, which was perfected into the human form they desired over a period of at least 200,000 years. This accords with what biologists say about the timeframe for the origin of the first "Eve" and, as will be seen later, it is no coincidence that the origin occurred in Southern Africa. Whatever version of the hominid was actually first utilised by the Annunaki, the reality from Sitchin's writings is that the creation of the human being desired by the Annunaki took many attempts, and only reached a successful conclusion with the version of the human that is known as Cro-Magnon, the early *homo sapiens sapiens*.

The success of this final version of the human relied on the Annunaki utilising their own genetic material to enable the new human form to be viable. Today's version of the human is a blend, or mix, a hybrid, of the Annunaki and *homo sapiens sapiens,* and started with the genetic manipulation of the hominoids or Neanderthals into the Cro-Magnons, who were attractive enough to cause the Annunaki to mate with them, and the offspring, then to continue mating over millennia to produce the version of humanity we see today.

Whist such writing is outside mainstream thinking, it is far from a crazy theory. This is because neither the arguments of Creationism or Darwinism do justice to the arrival of today's human being on Earth. As Robert Steven Thomas has said in his excellent book *Intelligent Intervention,* modern study in genetics and micro-biology have taught us that the formation of cells, their structure and the sophistication of their DNA model is a world that is far more intricate and complicated than was understood in Darwin's time. The leading book on the concepts that Robert Steven Thomas mentions is Michael J Behe's *Darwin's Black Box* which will be reviewed below.

Creationism is the belief that a Supreme Being created life and humans whole and complete, without any alterations or subsequent changes. For the purposes of the discussion of the creation of today's human, it can be seen from the sources quoted throughout this chapter that the concept of "creationism" doesn't quite fit. "Darwinism" is even further removed. Darwinism is the theory that simple forms of life develop into more complex forms of life by steady incremental progress powered by the concept of natural selection. The idea is that life forms continually upgrade themselves over time, with the relevant life form getting more and more sophisticated and strong by discarding the defective or outmoded parts. Unfortunately for Darwinism, there is a complete absence of transitional forms, and fossil records have never been able to verify Darwin's theory. There are plenty of fossil discoveries, just none that substantiate what Darwin proposed. We can find fossils of early and extinct primates, hominids, Neanderthals and *Homo Sapiens*, but no fossils of the transition linking ape and man. What we have been left with is *missing links*, not transitional species. Everything shows that changes happen abruptly, not gradually or by "slow steps" as Darwin maintained.

In fact, as is made abundantly clear from Lloyd Pye's book, not a single bone resembling human bones has ever been found before the arrival of the Cro-Magnon man some 120,000 years ago. What has been discovered are the bones of the hominoids and Neanderthals who were indigenous primates who developed on Earth alongside monkeys and apes and were not human. Lloyd Pye conclusively shows that hominoids are real and not a fantasy and, in fact, still exist on Earth, in areas of the planet not suitable for human existence. This then leads to the question of where the human came from if the bones/fossil records found from pre-Cro-Magnon times are not human. The answer according to Pye is "somewhere other than here", as 120,000 years (at most) is not enough time for the human being to have evolved according to Darwinian evolutionary theory. He concludes, based on his research and the research of others such as Zecharia Sitchin, that humans were developed/created on Earth by human-like "aliens" using a combination of genetic engineering and cross-breeding with native or indigenous earth-based beings who were available for genetic modification, in other words the hominoids.

The 5,000 contentious cuneiform clay tablets

It wasn't until the excavation of buried cities in the Mesopotamian region started a century and a half ago that the world had any serious idea about the civilisation which started around 6,000 years ago. Hundreds of thousands of clay tablets were discovered with cuneiform writing containing detailed records and documents of everything in the Sumerian society including politics, law, medicine, agriculture and trading, and the life and history of the people. But there were around 5,000 tablets which were contentious because they dealt with matters that were beyond the supposed everyday life of that time.

These tablets have been translated and show that society then was very different to what we would have imagined from that time. For instance they knew things that we didn't. They knew about astrophysics, and the characteristics of planets such as Neptune, Uranus and Pluto which were unknown to us in 1976 when Sitchin's first book of translations appeared (*The Twelfth Planet*). He stated that according to the translations, the planets Uranus and Neptune were a blue-green colour and were "watery twins" which was ridiculed by all the astronomers of that time, at the same time allowing the remainder of the book's contents to be conveniently written off. But when a satellite finally photographed these planets between 1986 and 1989 they discovered that what Sitchin had stated was, in fact, absolutely correct.

How could the Sumerians have possibly known this? The answer is that, as recounted in the tablets, they were given the information by beings from another planet, which they called the Anunnaki. The planet was called Nibiru, and was the "twelfth planet" of Sitchin's book title. They knew these astronomical details because they had flown past them many times from their home planet on route to Earth. They stated that there were twelve planets in our solar system, starting with the Sun, and moving outwards from the Sun, Mercury, Venus, Earth and its moon, Mars, Jupiter, Saturn, Uranus, Neptune and Pluto. This makes eleven, if you count the sun, moon and pluto as planets, leaving one more which we can't see: the Annunaki's home planet Nibiru, which is twice the size of Earth and has an elliptical, comet-like orbit around the sun of 3,600 years. The last time it came close to Earth on its elliptical orbit was allegedly around 2,200 years ago, in 200 BC.

The Annunaki were able to live on Earth in relative comfort, so their atmosphere must have been similar to Earth's. But why would they be interested in an undeveloped planet like Earth? Sitchin tells us that they came in search of gold to repair their atmosphere as gold flakes have excellent insulating and reflecting properties.

Sitchin produced a series of books which he called *The Earth Chronicles*, detailing why the Annunaki came to Earth, their existence pre- and post-flood, and the battles among the leaders which ended in them leaving Earth around 200 BC following a nuclear war they instigated. They mined for gold in Southern Africa, but this proved too arduous for the Annunaki and it was, therefore, agreed to create a slave to take over the hard mining work. This resulted in the genetic alteration of the hominoids which were native to the mining region, through gene splicing. The texts in the tablets give a perfect description of such a creature.

After many millennia of trial, a human version was created, the hairiness having been bred out, and these creatures were considered so attractive that the male Annunaki started to use the females as sex objects as well as using the male for the hard labour purpose. They had short lifespans however, and a lower IQ through the limitation placed on brain usage and brain size. Following the Great Flood, the earth was decimated, and the surviving Annunaki and their offspring spread out around the globe to settle in dry and habitable areas. They continued to propagate for several thousand years in their chosen areas until the war between the Annunaki leaders forced their departure leaving their hybrid offspring in charge of affairs following their departure.

Interestingly, in today's humans, certain blood groups do not mix with other blood groups. Negative blood types are said to occur where the humans come from a line of offspring directly descended from the Annunaki, whereas positive blood types descend from the hominoid hybridisation process. Some writers such as David Icke suggest that the negative blood groups have been carefully preserved over the centuries as the different blood types relate to elite breeding and that this has a relationship to today's ruling elite. We can see what happened to the control of Earth following the alleged departure of the Annunaki in the later chapter entitled "Global Control of Humanity", which references in particular David Icke's ground-breaking work *A Guide to the Global Conspiracy*.

The 23 chromosome pairs and gene splicing

But what is it that is so obviously different from a "natural" evolutionary process that would be seen if the Darwinian model of evolution had been followed in the human? Lloyd Pye at page 292 onwards in *Everything You Know is Wrong* discusses this issue, stating that despite huge segments of genetic code that match perfectly, humans and apes such as gorillas, chimpanzees and orang-utans possess some highly unusual differences. The most significant of these is the reduction of the great apes 24-chromosome pairs down to the 23 that humans possess. Pye shows how genetic engineering would not allow the removal of an entire chromosome as this would do too much damage; so the answer is 'gene splicing', whereby two chromosomes are spliced together to create one in the human. A key difference is the spliced chromosome, an impossible feat to have naturally occurred. But there are others, such as the fact that human chromosomes have several subtle inversions, and many key segments reside in non-matching locations, which are strange "rearrangements".

The Annunaki did not get everything perfectly right either, as Pye points out. Humans have gene-based disorders such as cystic fibrosis and diabetes which are almost never seen in the animal kingdom as these are naturally eradicated, as well as trinucleotide repeat diseases that have never been seen in any other species studied by geneticists. It makes no evolutionary sense that we carry over 4,000 built-in genetic disorders while our closest genetic relatives carry almost none. Apes don't continue to produce defective specimens like humans do. For genetic disorders to become permanent fixtures in the gene pool of a species that are passed on to the next generation, they have to have been *put into* the DNA code itself. This is not a natural process. And, as Pye observes, wouldn't a Supreme Creator have left the human in better physical shape than apes, monkeys, hominoids and other 'lower' creatures?

It is a reasonable deduction, therefore, that the current human being was created and did not evolve naturally.

Interestingly, Lloyd Pye points out that whilst Darwin's theory of evolution is appropriate at the micro-evolution level, as birds, animals and so on will develop through adaptive modification depending on circumstances, one type of creation does not suddenly turn into another. At page 19 of *Everything You Know is Wrong*, he states: "Sea worms did not and do not become fishes, fishes did not and do not become amphibians, amphibians did not and do not become mammals. In every case, the differences between critical body parts and functions (internal organs, digestive tracts, reproductive systems, etc.) are so vast that transition from one to another would require dramatic changes that would be easily discernible in the fossil record. What the fossils record actually reveals is that every class, order, family, genus, or species simply *appears*, fully formed and ready to 'eat, survive, reproduce'."

The black box – how life works

One of the biggest problems for Darwinism is the recent discoveries and developments in biochemistry. These have been outlined by Michael Behe in *Darwin's Black Box*. Biochemistry is the study of the very basis of life: the molecules that make up cells and tissues that catalyse the chemical reactions of digestion, photosynthesis, immunity and more. Behe intends that the term 'biochemistry' should include all sciences that investigate life at the molecular level. Behe shows that the history of biology is a series of "black boxes" which, ultimately, lead to the conclusion that the idea of Darwinian evolution is unworkable, at least at the molecular level. Nearly twenty years after the initial release of his ground-breaking work (*Darwin's Black Box*), he refined the position established in that book in his 2007 publication *The Edge of Evolution* by showing where the boundaries of "evolution" are, and where random mutation and natural selection are not possible. Both books are essential reading on the topic of intelligent design. But *Darwin's Black Box* is where the key arguments are set out, and that book will be briefly reviewed below.

A "black box" is a term for a device that does something but which has inner workings that are difficult to fathom out. For instance, we all use computers, but few of us know how they actually work. The ultimate "black box" in biological terms is the "cellular black box". It took a long while for this "box" to be opened and properly understood (well after the time of Darwin), because many small, critical details of cell structure cannot be seen with a light microscope, as the wavelength of visible light is only around one-tenth the diameter of a bacterial cell. It took the discovery of the electron and the invention of the electron microscope for cellular diagnosis to advance, because the wavelength of the electron is shorter than the wavelength of visible light, which meant that smaller objects could be resolved by "illumination" with electrons. It transpired that the cell was very complex.

This discovery allowed biologists to open the "greatest black box of all" and discover *how life works*. It was the greatest black box because it was the *last* black box, beyond which we cannot go, and when opened it revealed molecules, the "bedrock of nature". As Behe puts it: "Unlike earlier scientists who looked at a fish or a heart or a cell and wondered what it was and what made it work, modern scientists are satisfied that the actions of proteins and other molecules are sufficient explanations for the basis of life. From Aristotle to modern biochemistry, one layer after another has been peeled away until the cell – Darwin's black box – stands open." Behe's book demonstrates, through examples relating to the working of the human body, that the Darwinian theory of evolution is unable to account for the molecular structure of life, because it is unable to answer the question of how complex biological systems could be *gradually produced*.

Darwin actually postulated in his book *Origin of Species* that his theory of gradual evolution by natural selection would break down if "it could be demonstrated that any complex organ existed which could not possibly have been formed by numerous, successive, slight modifications".

Irreducible complexity

Michael Behe shows in *Darwin's Black Box* that the criterion for failure of Darwin's theory has been met.

This is shown through a system that is irreducibly complex, by which he means "a single system composed of well-matched, interacting parts that contribute to the basic function, wherein the removal of any one of the parts causes the system to effectively cease to function". Thus, such a system could not have evolved slowly, piece by piece. Behe uses the concept of irreducible complexity to support his belief in intelligent design of life forms. This is because of the fact that irreducible complexity is present in the creation of the first single-cell animals, which makes random creation essentially impossible.

Right at the beginning of creation, the single-celled amoeba, whilst appearing to be a simple organism, is, in fact, a marvel of creation so complex that it has all the integral parts for sustaining life and, if any single part was ill-formed or missing, the remaining parts would all be useless and unable to reproduce and, above all, none of the individual parts is capable of replicating or sustaining themselves independently.

Behe points out that none of the top scientists, academies, institutes or universities can give a detailed account of how the "cilium, or vision, or blood clotting, or any complex biological system might have developed in a Darwinian fashion". So how did they occur? If they were not put together gradually, then they must have been put together quickly or even suddenly.

Behe's conclusion is that many biochemical systems were *designed*, in fact, *planned*. By design he means the *purposeful arrangement of parts*. Behe states: "The designer knew what the systems would look like when they were completed, then took steps to bring the systems about. Life on earth at its most fundamental level, in its most critical components, is the product of intelligent activity".

One can infer design because a number of separate components are ordered to accomplish a purpose that none of the components could do by itself. By intelligent design he meant design that goes beyond the laws of nature.

It is not a distant "jump" to conclude that design of the "cellular black box", the molecular foundation of life, was likely to be of extraterrestrial origin. Why should people continue to deny the existence of design when Behe's book makes the case so powerfully? What is so worrying about the possibility of "intelligent design"? This seems to be an area which will be less threatening once the reality of extraterrestrial existence is revealed, and which will ultimately fall within the realm of exopolitics (a field of activity discussed later when reviewing Michael Salla's *Exposing US Government Policies on Extraterrestrial Life*.

DNA

Further, DNA is the mechanism enabling reproduction of all forms of life on earth. Graham Hancock sets out in immense detail in *Supernatural* the fact that the make-up of DNA of is too complex to have materialized by chance. The four-letter DNA 'language', together with the 20-letter protein 'language' and the translation mechanism that links the two were far more likely to have been engineered or 'created', before it arrived on earth. Crick (and the astrophysicist Sir Fred Hoyle) felt it must have been created by intelligent aliens billions of years ago. It seems that for the first 600 million years of earth's 4.5 billion years of existence it was simply a molten lava fireball, but by 3.9 billion years ago cooling was sufficiently advanced to produce a thin outer crust of solid rock. It is supposed that around the same time pools of water enriched with minerals began to take shape beneath an atmosphere of simple gases.

In this water, primitive life forms appeared suddenly. Crick felt that it was essentially impossible for instant life to form in this way without assistance. These early life forms contained the DNA that still makes life today. Over the billions of years that followed, life forms evolved and the messages in the DNA sat there waiting for a sufficiently evolved creature to be able to make use of the information coded in the DNA by intelligent alien entities. If the creator didn't know what form of creature would ultimately evolve, the only place to hide the secrets of creation would be in the DNA itself, waiting for a sufficiently intelligent creature to unravel the codes. Crick speculated that DNA was preserved in earth creatures by alien entities to ensure its survival from a catastrophic event elsewhere in the universe. Our DNA is certainly a thing of wonder. DNA is a crystalline substance with a shape that makes it a perfect receiver-transmitter. The membrane of every human cell is a liquid crystal, of which we have trillions in our bodies. These cells could be called the computer hard drive of the body and the software is the crystalline DNA holding the genetic memory. We have some 120 billion miles of DNA in our bodies and it can store more than a hundred trillion times more information than any device that human science can construct.

Water – an extraterrestrial gift?

And following on from the above mention of water, in his book *The Hidden Messages in Water*, Dr Masaru Emoto makes an interesting observation that water may even be 'not of this world'. Water may well have come to earth from elsewhere in the cosmos at the time earth was formed 4.6 billion years ago, or possibly (as propounded by Professor Frank of the University of Iowa), through an onslaught of lumps of ice which hit earth from outer space (and continues even today). Dr Emoto's findings, together with those of Professor Frank, are lengthy, but NASA says the theories may have credibility. It is an absolute fact that there can be no life without water, and if we accept that water, the source of all life, was sent from outer space, then it follows that all life, including human existence, is alien to this planet. It seems that water emanates from outer space, arrives as lumps of ice on earth, and then becomes clouds and eventually falls to earth as rain or snow. The water then washes the mountains, seeps into the ground, becoming rich in minerals, and then rises to the surface again. The water is then carried from the mountains to the rivers and on to the ocean. Movement is the key.

As with our DNA, water carries information. It seems that when the water left outer space and fell to earth, it carried with it the program needed for the development of life. Water records information and then, while circulating throughout the earth, distributes information. The information distributed is the information about life itself, how it works. Dr Emoto deciphered this information through the observation of ice crystals.

Genetic Engineering and Spiritual Animation

Sitchin stated, in his *Earth Chronicles* series, that the (non-human) occupants of Earth at that time, who he says had come to exploit Earth's mineral resources, found the work took too much of a toll on their bodies. As mentioned earlier, Earth's occupants therefore created a new creature from a human format that already existed on Earth in plentiful numbers that was capable of performing the difficult physical labour, and take the strain away from them. Thus was created today's human being. The concept of genetic engineering is no longer a far-fetched concept, as today scientists are able to clone and carry out hybridisation work in this very fashion. Note for instance in 2008 the passing of a new law by the UK Parliament known as the Human Fertilisation Embryology Act which allows scientists to experiment with cloning techniques, legalising the creation of a variety of hybrids, including an animal egg fertilised by a human sperm, 'cybrids' in which a human nucleus is implanted into an animal cell, and 'chimeras' in which human cells are mixed with animal embryos. Sitchin, and others, tell us that the rulers of Earth animated this new creature with a spirit, or 'consciousness', so that it had awareness, understanding, intelligence and personality. Otherwise the creature would be no more than a reactive animal, and of no real use. So their genetic engineering ensured that spiritual beings were permanently attached to human bodies to animate the bodies. This would make them intelligent enough to perform their required tasks.

But this spiritual animation did not mean that the humans were allowed to experience happiness. Quite the contrary. As Bramley so eloquently explains throughout the book, methods of control were carefully introduced to ensure that humans remained controlled, and this continues to this day. As the translation of the cuneiform tablets states, "There must be no rejoicing among them".

Bramley states that the Controllers clearly did not want the newly-created mankind to begin to travel down the road to spiritual enlightenment. He states that the reason for ensuring that the human 'spirit' is not awakened is because obviously the rulers wanted slaves, not free beings. Physical threats and the creation of fear are not going to have the same effect if humans start to understand that they are spiritually immortal, and the death of the physical body is not the end. Or worse still, during the spirit's occupation of the physical body they discover that they could abandon that body at will and become detached when they desired. That understanding and awareness had to be blocked, as will be discussed further in the later chapter entitled "Global Control of Humanity".

The Moon: an alien creation to support the development of life?

"A super-intelligence is the only good explanation for the origin of life and the complexity of nature".

So said Professor Anthony Flew in December 2004 at the age of 81, having spent the years until he reached his eighties as a leading exponent of atheism and logical thinking.

Professor Flew had revised his opinion following an analysis of new information which led him to conclude that science appears to have proven the existence of God. In particular, this conclusion was reached following recent investigations into DNA by biologistswhich has shown, that an unbelievable complexity of arrangements are needed to produce life, leading to the conclusion that intelligence must have been involved.

But the authors Christopher Knight and Alan Butler in their book *Who Built the Moon?* have taken this idea one step further: that the moon is very much part of intelligent creation, and is integral to the creation of life on Earth. There is something very odd about the Moon for sure. The facts we know about the Moon leave many questions unanswered.

These are systematically answered by Messrs Knight and Butler, who conclude: Both the Moon and the Earth are approximately 4.6 billion years old; the Moon was manufactured from lighter materials taken from a young Earth; the Moon was made as an incubator to foster life on Earth; and the manufacturer of the Moon seeded life on Earth. Sounds unbelievable? After reading *Who Built the Moon?* one is left in no doubt that the scientific facts cannot be disputed: The Moon was built, not created like stars and other celestial objects.

To start with, the Moon is bigger than it should be, older than it should be and much lighter in mass than it should be. Moreover, scientists consider that the Moon may have an unusually light core, or perhaps even no core at all. In other words, the Moon is hollow. When the Apollo 13 craft crashed its booster into the Moon it is reported that the Moon "rang like a bell", and that the whole structure of the Moon 'wobbled' in a precise way, 'almost as though it had gigantic hydraulic damper struts inside it'. Professor Carl Sagan made it clear that a natural satellite (which the Moon has always been thought to be) cannot be a hollow object. The fact is that solid objects do not ring like a bell, but hollow ones do. So if the Moon *is* hollow, it leaves only one conclusion: someone or something manufactured it.

The Knight and Butler book is full of incredible facts, and it is worth mentioning a few more here:

The Moon is one 400th the size of the Sun: the Moon is 400 times closer to the Earth than the Sun; the Moon is rotating at a rate of 400km per earth day; the earth is rotating at 40,000km per day, and the Moon is turning at precisely 100 times less; the Moon always faces the Earth as it travels on its orbit around our planet and yet the average distance is such that the equatorial rotation speed is precisely one per cent of an Earth day; the Earth has 366 revolutions for each orbit around the Sun, and the Moon is 366 per cent smaller than the Earth; and so it goes on. Surely this can't be random?!

The real fact is that without the Moon in its precise position in relation to the Earth, life on Earth would not be possible. The Moon is Earth's life support machine, and that is why Earth has spawned life and other planets nearby have not. There are two other factors which are unique to Earth: its tilt and plate tectonics.

Earth's tilt

If the Earth did not tilt, life would be almost impossible across most of the planet but, fortunately, the Earth is at an angle of around 22.5 degrees relative to the equator of the Sun, and this angle is maintained by the Moon, which acts as a giant planetary stabiliser.

The tilt angle of 22.5 degrees ensures that most parts of the Earth's surface get a fair share of warmth throughout the year. This means that the vast majority of water on the surface of the planet remains in a liquid state. This is critical, because life cannot exist without water. So, as Messrs Knight and Butler say, the Earth is extremely well balanced. It also means that there are few parts of the globe that cannot support human life, as the temperature range is very narrow. No other known planet has such a narrow temperature band. Most planets are either too cold or too hot to allow human life. With only a small change in the overall temperature of the planet, or an alteration of seasonal patterns (made possible by the tilt), the nature of the water on our planet would change and be unusable.

The point of all this, though, is that it has been vital for our existence that the tilt of the Earth has been maintained at around 22.5 degrees for an extremely long period of time, something that would not have happened without the Moon. The extremely large Moon has ensured that the tilt does not vary relative to the Sun, and has acted as a vital stabilising influence. Other similar planets, such as Mars, Mercury and Venus, have wildly varying tilts. For instance, Mars varies between 0 degrees and 60 degrees as it doesn't have a suitable sized Moon to stabilise it. It has been projected that Earth would have an even wilder tilt, between 0 degrees and 85 degrees were it not for the Moon. So life is entirely dependent on our Moon. It has also slowed Earth's spin, allowing stability for life to develop.

Earth's plate tectonics

And so to Earth's 'plate tectonics', which is a truly awesome reality. Plate tectonics do not take place on any other of our neighbouring planets. The book *Who Built the Moon?* puts forward the entirely rational concept that the Moon was built from Earth materials based on findings relating to samples from the Moon's materials. This has allowed life to develop on Earth.

Any intelligent guiding force which desired that life would be nurtured on planet Earth would know that the planet would need to have a loosened surface and, in view of the fact that a regulator for life on planet Earth would be needed, the obvious solution would be to take surface material from Earth and use it to manufacture the regulator. This would reduce the tendency of the surface to form one continuous crust and would allow movement within the crust itself. On Earth, the elements most crucial to life, carbon and water, are continuously recycled due to plate tectonics which, as mentioned above, is the slow, steady movement of great pieces of semi-rigid crust over the more plastic layer of the upper mantle.

So to make the whole plate tectonic process work, the answer was to move seventy-four quintillion tonnes of material from the surface of the young Earth and manufacture the required regulator of Earth (in other words, the Moon) from it. To meet the requirements, the regulator needed to have a mass of just 1.234 per cent of the altered planet, but still be 27.322 per cent of the size of the planet. This meant it would have to be made with minimal heavy elements, such as iron, and be essentially hollow. And this is what was, indeed, manufactured, allowing life to develop on Earth. A fantastic reality.

Knight and Butler quote Nick Hoffman, an acclaimed expert on the terrestrial planets within our solar system, who has suggested that the removal of the material that went to make the Moon may have triggered plate tectonics by creating space for the planet's skin to shift, allowing continental drift to take place, even though this would have occasional catastrophic consequences (as can be seen from the information about cataclysms in the later chapter entitled "They Left Signs That They Were Here"). If the materials were returned to Earth it would 'fill the ocean basins with wall-to-wall continent' and without plate tectonics Earth would be a water world, with only the tips of extremely high mountain ranges poking out above the surface of the water. But plate tectonics allowed life to form on the planet. One hopes that after reading the fascinating work and factual materials contained within *Who Built the Moon?* disbelievers may be propelled to follow the likes of Professor Anthony Flew and start to believe that an intelligent force was at work in the creation of life.

Phobos – an ET space station?

But it is not just Earth's satellite that may have been artificially created. The moons of Mars, 'Phobos' and 'Deimos', may also be artificial satellite creations. Sagan and Shklovsky, eminent astronomers, stated that a natural satellite cannot be a hollow object. And that is what Phobos and possibly Deimos appear to be if data from the European Space Agency is correct. In 2008, a flyby probe produced data from which the European Space Agency concluded that "the interior of Phobos likely contains large voids". Indeed, in 1959, Shklovsky went further and suggested that Phobos is in fact an ET space station. Shklovsky's claim may have some substance as, in 1991, the Russian author Paul Stonehill was allegedly shown a photograph by Russian Cosmonaut Marina Popovich, taken from the unmanned probe Phobos 2 before its demise. Incredibly, this depicted a gigantic cylindrical object approximately 15 miles long. This is detailed in Stonehill's book *Russia's Roswell Incident*.

Moreover, it has been postulated that Mars itself may have been a base for alien existence many years ago due to the apparent artificial structures photographed on its surface.

Mars – life below the surface?

In Dirk Schulze-Makuch and David Darling's 2010 book *We Are Not Alone*, one conclusion was that earlier in solar system history, Mars was likely more suitable than Earth as an incubator of life because our own world took much longer to cool down from its hot and fiery beginnings. Could this have been why life existed on Mars at one time, now extinct, but which may have "moved on" to Earth at some point as Mars became increasingly uninhabitable as a base? Mars is not readily habitable on the surface these days because of low atmospheric pressure and low surface temperature, lethal solar radiation and air that is unbreathable, but could life still be possible below the surface?

Recent observations have pointed to an abundance of water ice just below the surface, and even surface water and falling snow. Dirk Schulze-Makuch and David Darling point out that research in the year 2000 identified more than 120 locations where it appears that water had seeped into freshly-cut gullies and gaps on the Martian surface. It is now known that water is present on Mars in much greater quantities than had earlier been considered possible and, if heat were present, one would have the required environment like on Earth for nurturing life. This seems likely in view of the 2004 detection of methane gas in the Martian atmosphere, and then, in 2005, the detection of enormous quantities of formaldehyde, which would favour microbial life.

Space engineering

Returning to the concept of artificially created objects in space like the Moon, perhaps the idea of massive solar construction projects is not so far-fetched if the work of Norman Bergrun in his book *Ringmakers of Saturn* has any validity. Bergrun's book is quoted by Joseph P.Farrell in his own book *Covert Wars and the Clash of Civilisations*, where it is stated that Bergrun analysed pictures from NASA's Voyagers I and II when they did a flyby of Saturn in 1980 and 1981 respectively. Bergrun's conclusion was that the pictures show "immensely large, enormously powerful extraterrestrial space vehicles located in the vicinity of Saturn and its moons" which appeared to be engaged in massive engineering and construction projects, including building Saturn's rings. These crafts were estimated to be about 1,730 miles wide, which would make them space vehicles beyond our wildest imagination, and clearly capable of constructing moons.

But before we consider the impact on humanity of alien contact, any review of the literature on extraterrestrial life and its relevance to humanity has to have a starting point. And the starting point is of course: How did the extraterrestrial Controller race arrive (and how do extraterrestrials of all sorts continue to arrive) on Earth? The laughably simple answer is that it is through the use of a flying craft (the UFO as we now know it).

PART 3

Flying In (and Out)

PART 3

Phasing In and Out

The appearance of the UFO

As mentioned above, the laughably simple answer to how the extraterrestrial Controller race arrived was through the use of a flying craft. Throughout history and the existence of civilisations on Earth, there have been many reports of flying crafts which are unknown to that civilisation.

The ancient Mesopotamians and the ancient Egyptians both claimed to be living under the rule of human-like extraterrestrial "gods". The Egyptians wrote that their "gods" travelled into the heavens in flying "boats". Some "gods" such as the God of Zoroaster, Ahura Mazda, flew around in a flying object that had landing pads. The Apocrypha books not included in the Bible suggests that the Star of Bethlehem may have been a UFO. The Book of Revelation in the Bible details in the famous passage at 4:1-6 the experience of John being taken through the door of some sort of craft and finding himself face to face with the occupants. Reports of craft were plentiful during time of disease and pestilence through the centuries.

In spite of all these reports, claims and writings, officialdom only began to take note of aerial phenomena at the time of the First World War when strange lights and craft were reported. Since then millions of sightings have been reported across the world. Yet the official position around the world is still that extraterrestrial craft do not exist. This goes back to a specific moment in history: the Roswell crash in July 1947. Within hours of the 1947 crash at Roswell, New Mexico, USA which was reported as the crash of an alien craft, US General Twining apparently designated the visitors (being the craft and its occupants) as 'enemy aliens'. It appears this has remained official US policy ever since. This policy was followed Governments around the world.

For a comprehensive high level view on the reality of extraterrestrial existence, for those of us who are not dedicated researchers, two books represent essential reading. These are the 2010 book by Len Kasten entitled *The Secret History of Extraterrestrials* and the 2009 book by Michael Salla entitled *Exposing US Government Policies on Extraterrestrial Life*. These two books alone lead one to the conclusion that extraterrestrial existence is simply being purposely hidden from view by the authorities, but together with the review of other key literature in this book the reader should be able to gather a good level of knowledge about what appears to be going on behind (for the most part) closed doors.

There is simply too much evidence. It is merely being contained and not released to a wider audience for fear of the panic it would probably rightly cause.

Timothy Good in *Need To Know* has taken UFO research to a point where it is frankly impossible to say that extraterrestrial crafts do not enter Earth's atmosphere. For instance, as long ago as 1947 the US commanding general of Air Material Command wrote to the intelligence chief of the Army Air Forces that "the phenomenon reported is something real and not visionary or fictitious". The following year a top secret US document concluded that the UFOs were interplanetary in origin and were a product of high technical skill which cannot be credited to any presently known culture on Earth.

Many of the sightings occur where nuclear and other missile bases are situated, suggesting the craft are observing, and occasionally even disabling the weapon systems. The famous Rendlesham Forest incident at the USAF base in Suffolk, UK, (the "UK's Roswell") suggests that the craft are surveying the stockpiles and capabilities of our weapons. Sightings of UFOs were commonly reported during ICBM test launches, maneuvering around them at incredible speeds and sometimes disabling the dummy nuclear warheads. One of the most extraordinary incidents was reported by Linda Moulton Howe in *Glimpses of Other Realities* concerning the almost impossible feat carried out by UFOs of shutting down fifteen nuclear missiles on March 16, 1967 at the Malmstrom Air Force Base.

The 1947 Roswell occurrence (which is referenced later) did not bring UFOs to the attention of the wider public because at that time it was covered up, but the 1952 sightings of flying saucers over the Washington DC area generated headlines in newspapers throughout the world. It was difficult for the Government to deny the occurrences as so many people had witnessed them, and it led the then President Truman to summon assistance from the Air Force in dealing with growing public concern.

During the Washington sightings it was also reported that 'gigantic' flying objects were tracked orbiting Earth. At one point they descended to around 79,000 feet and caused pilots who saw them to admit being scared of the massive crafts. They appeared thankful that they couldn't get close even to the height to which they had descended.

Gordon Cooper, a test pilot and one of the original Mercury 7 astronauts, wrote in his autobiographical book *Leap of Faith* that in 1951 (the year before the Washington sightings) while he was stationed in Germany he witnessed his first UFO flights. He states that he and other pilots were scrambled in F-86's to intercept the "bogies". However, the F-86's reached their maximum ceiling of 45,000 feet and were still way below the metallic silver saucer-shaped objects. In fact, Cooper reports that the metallic craft were so high that it was impossible to get any idea of their true size. He states that over a period of days the craft appeared in groups of four, and at other times as many as sixteen. They moved fast, slow, and even stopped in the air, and flew right over the air base in Germany. Eventually the pilots stopped going up after the craft, as they simply flew too high and fast. Cooper states that the pilots simply looked to the sky at the craft through binoculars "in awe at the speedy saucers". There were many reported incidents (Good's book *Need to Know* covers this extensively) of jets failing and falling out of the sky when they attempted to attack UFOs, with the pilots losing their lives.

The reports of UFO's were so worrying that a US Government panel which was set up in 1953 concluded that "the continued emphasis on the reporting of these phenomena does, in these parlous times, result in a threat to the orderly functioning of the protective organs of the body politic". The resulting ban on information has never been seriously lifted since then, and the secrecy policy was extended to the Soviet Union, Great Britain and France at a secret meeting in Geneva in 1955. China is also known to have extensively gathered UFO information though not much has leaked out.

Attempts have been made to lift the ban. For instance in the early 1960's two former CIA officials who sat on the National Investigations Committee on Aerial Phenomena (NICAP) wrote that information on UFOs was dangerous to withhold. They stated that the craft were "under intelligent control. The speeds, maneuvers and technical evidence prove them to be superior to any aircraft or space devices now produced on Earth. These UFOs are interplanetary devices systematically observing the Earth, either manned or under remote control, or both."

John Podesta, President Clinton's Chief of Staff, stated in 2002 that it was time for the American people to hear the truth about UFO's because (amongst other things) "they can handle it". However, with his vast knowledge of UFOs gained through his UFO research, Timothy Good concluded at the end of his 2006 book *Need to Know* that disclosure would be destabilizing. Gradual disclosure, which is the planned agenda, is a wiser course.

To back this opinion up, Good's research into the Brazilian experiences of UFO sightings show that they engendered widespread anxiety and alarm. Moreover, the reported incidents around ten years ago of alien craft in Ireland posing a threat to commercial airline safety while the airliners were on landing approach would be likely to cause considerable public concern if the authorities did not keep it under wraps. Perhaps criticism of Government censorship and suppression of

UFO reports should be better understood in this context. The UFO subject, according to one of Good's sources in Washington, ranks as the most sensitive in the US Government.

In 1965, a former officer of the French secret service revealed to a French journalist that the Russian and American secret services had collaborated on the problem and had concluded: "The flying saucers exist, their source is extraterrestrial, and the future – relatively quite soon – should permit confirmation of this statement." No confirmation has ever been received.

Even the official spokesman for the UK Government in the early 90's, Nick Pope, alluded to the fact that there may be a lot more information available than the UK Ministry of Defence has in its files. Whilst Pope's book *Open Skies, Closed Minds* was a very readable summary of the UFO phenomenon at its time of writing (1994), it is apparent that anything remotely difficult to explain away is conveniently buried under the category of "Unexplained", or "Unknown", when type, origin and occupants of a craft are not ascertainable. Certainly some of the cases he highlighted defy conventional explanation and may be subject to either national security implications or covered by secrecy for other reasons.

Gordon Cooper mentioned in *Leap of Faith* that between 1948 and 1969 the United States Air Force investigated 12,618 reports of UFO sightings under a program known as Project Blue Book. All but 701 of these sightings were dismissed as balloons, satellites, aircraft, lightning, reflections, astronomical objects such as stars or planets, or outright hoaxes. However, the remaining 701 sightings were classified as "unexplained", and have never since been explained. But when

Project Blue Book was closed in 1969, the final report provided a complete whitewash of the UFO phenomena, stating that there had been no evidence indicating that sightings categorised as "unidentified" were "extraterrestrial" vehicles. Moreover, no UFO reported, investigated or evaluated had ever given an indication of threat to national security. UFO researchers in the USA often find that information is not made available under the Freedom of Information legislation due to national security implications. Does someone know more than is being revealed?

In their book *Real Aliens, Space Beings, and Creatures from Other Worlds*, Brad and Sherry Steiger, two of the world's leading authorities on aliens and UFOs, detailed types of aliens that are known to this world. They range from attractive tall human-looking blondes known as "Blonde Nordics", small creatures with large heads and wide black eyes known as "Grays", insect-like creatures resembling praying mantises, to serpent-like reptilian creatures. More detail on these will be given in later chapters. The Steigers' book covered a vast area of alien activity, but one thing they noted was that a non-terrestrial source is definitely interested in what is taking place on planet earth.

The Steigers stated that, for at least the past sixty years, there has been a space program originating from somewhere other than the known terrestrial space centres. It has been placing satellites in orbit around our planet and transmitting bizarre messages, and this has for the most part been successfully hidden from the general public. It has been theorised by some researchers that these "unknown orbiting vehicles" are probably launched from bases on either the Moon or

Mars. The Steigers reported that in 1960, three mystery objects were tracked by NORAD, the anti-missile defense radar network established by the Pentagon, and were also tracked by NASA. Following a leak from NORAD, it was revealed that the three mystery satellites were massive structures. These three mystery satellites still remain in orbit and, consequently, the International Satellite Authority (centred in France) which grants official designations to all satellites launched by any nation on Earth, had no choice but to add the three unknowns to their list and provide them with catalogue numbers.

Something is 'out there' and it's not from Earth.

Traversing space

Bruce Cathie, a former pilot and a scientific researcher, explained a little of what he had surmised about UFO propulsion in a paper he wrote called *Mathematics of the World Grid*. He discovered that gravity and light are reciprocals of each other and UFOs manipulate these waveforms and frequencies to travel. He said that the only way to traverse the vast distances of space is to possess the means of manipulating, or altering, the very structure of space itself, altering the space-time geometric matrix, which to us provides the illusion of form and distance. The method of achieving this lies in the altering of frequencies controlling the matter-antimatter cycles which govern our awareness, or perception, of position in the space-time structure. Cathie says time itself is a geometric, just as Einstein had postulated; if time can be altered, "then the whole universe is waiting for us to come and explore its nook and crannies". In the blink of an eye we could cross colossal distances - for distance is illusion.

The only thing that keeps places apart in space is *time*. If it were possible to move from one position to another in space, in an infinitely small amount of time, or 'zero time', then both the positions would coexist, according to our awareness. By speeding up the geometric of time, we would be able to bring distant places within close proximity. This may be the secret of the UFOs: They travel by means of altering the spatial dimensions around them and repositioning in space-time. Certainly this is one explanation given by some contactees.

Otis T Carr designed a civilian spacecraft which created an entirely new gravitational field inside the craft. This effectively created a zero mass environment inside his craft that would suspend the normal laws of inertia. The zero mass environment would enable the spacecraft to achieve light speed velocity. It would also enable occupants to withstand tremendous accelerations and changes in direction without being pulverized by immense g-forces inside the craft.

The theories of the likes of Cathie and Carr about travel faster than the speed of light go a long way to challenging the mainstream scientific view that the speed of light represents an insurmountable obstacle to the physical presence of extraterrestrial craft and visitors.

It would be reasonable to assume that this information is well known to Government scientists who have spent time back-engineering crashed crafts and that, in the years since the first craft crashed, UFO propulsion methods have been replicated by Government scientists and we can now travel through space and time in just the same way as the extraterrestrials. However this is not information that would readily be shared with the public.

Len Kasten reports in his *Secret History of Extra-terrrestrials* that informants under deep cover in high places state that we now have antigravity craft capable of routine intra-solar system travel and deep-space ships that use even more sophisticated propulsion systems such as tachyon and antimatter engines, and that we have now learnt how to use 'stargates'.

In her 2006 book entitled *UFOs: How Does One Speak to a Ball of Light?*, Paola Leopizzi Harris reproduces a series of interviews she undertook with leading participants in the strange world of alien activity. One of the interviews was with Al Bielek who took part in the famous "Philadelphia Experiment" where the USS Eldridge was transported, incredibly, from the Philadelphia Naval Yard to Norfolk Virginia and back again in a matter of seconds, in what can only be described as "time travel".

The ship had experimental electronic equipment on board with the ability to generate massive electromagnetic fields and bend light and radio waves around the ship. When the equipment was powered up, the ship became not only radar invisible, but invisible to the eye as well. A greenish fog enveloped the ship and it disappeared and then reappeared in the Naval Yard. What appears to have happened according to Al Bielek was that a dimensional porthole had been opened which allowed the ship to 'jump' from one place to another in the blink of an eye. Bielek said that this produced a sort of 'wormhole' effect, actually causing a breach in space-time allowing matter to jump through space-time. As a side effect Bielek said that alien craft from another dimension or universe actually entered our universe during this experiment.

Bielek explained that these dimensional portholes have been called 'wormholes', which are hyperspace tunnels through space-time connecting together either remote regions within our universe or two different universes, and possibly even different dimensions and times. Space travellers would enter one side of the tunnel and exit out the other, passing through the tunnel to get from place to place. This was how space travel was undertaken. These tunnels are also known as 'stargates'.

To understand how far technology has developed, in May 2010 it was reported in the MUFON (Mutual UFO Network) journal that Ben Rich, the former head of Lockheed's secret research and development unit, had stated before he died in 1995 that the US military now has the technology to travel into distant space, to the stars, but that these technologies are locked up in hidden "black" projects, and are unlikely ever to be used for humanity's benefit. As Colonel Corso stated in his extraordinary book of revelations about the Roswell crash entitled *The Day After Roswell* (which is reviewed later), it appears that the Roswell crashes were a primary source in anti-gravitational research and development. This research and development work was undertaken by corporations rather than the government to ensure that information about the development programs did not get leaked to the public through the work of public information researchers utilising the Freedom of Information laws. If information that has nevertheless leaked from informants is valid, there would now appear to be both man-made UFO's as well as the original extraterrestrial UFO's.

UFO's are real

The late William Cooper was one of the many highly placed military men who said that extraterrestrial presence was real. Before he stepped out to reveal his information to the general public, information about alien presence from credible sources had largely been kept secret. However, on 2nd July 1989 he made a speech at the MUFON Symposium in Las Vegas, Nevada which revealed alleged secrets of the US Government relating to extraterrestrial entities that he said he had access to as a former naval intelligence officer during the years 1970-1973. The speech was entitled "A Covenant with Death" because of what he perceived would happen to him following the revelation of the information contained in the speech. The speech can be accessed on the internet.

The contents of his speech formed the basis for his book *Behold a Pale Horse* and, following his July 1989 speech, he became a regular speaker about UFO and paranormal activities. The information he presented back in the late eighties onwards certainly appeared fantastic and hard-to-believe at that time, and caused a great stir, although the slow process of UFO and alien revelation is now desensitizing us.

Set out below is a summary of his speech. The essence of his speech was that after World War II the

US Government was confronted with a series of 'events' which changed its future beyond anything anyone could have imagined. And because America was the most powerful and richest nation on the planet, it affected the future of humanity too.

Cooper stated that these 'events' were that between January 1946 and December 1952 at least 16 crashed or downed alien craft, 65 bodies, and one live alien were recovered from within the borders of the United States. Additionally, it was discovered that some of these crafts were storing human body parts in the crafts. The live alien that had been found wandering in the desert from the crash at Roswell was named EBE, short for 'Extraterrestrial Biological Entity'.

Cooper went on to say that plans were formulated (under immense secrecy for fear of scaring the population at large and causing the collapse of religion, society and the economy) to defend the Earth in case of invasion by aliens. It was decided that an outside group was necessary to coordinate and control international efforts in order to hide the secret from normal scrutiny of governments by the press. The result was the formation of a secret ruling body which became known as the Bilderberg Group, after the place where they first met, the Bilderberg Hotel. The headquarters of this Group is in Geneva, Switzerland. The 'Bilderbergers' evolved into a secret world government that now controls everything. (For further information on global control, see the later chapter entitled "Global Control of Humanity".)

Cooper said meetings were subsequently held during the 1950's with members of the alien group who had crashed their craft in the Arizona desert in the 1940's. A treaty was entered into which stated that the aliens would not interfere in human affairs and humans would not interfere in alien affairs. The deal was that the aliens would furnish the US Government with advanced technology and would help in technological development. No treaties would be made with any other nation on Earth. In return the aliens were permitted to abduct humans on a limited and periodic basis for the purpose of medical examination and monitoring, with the stipulation that the humans would not be harmed, would be returned to their point of abduction, would have no memory of the event, and that the aliens would furnish a Government Group known as MJ12 with a list of all human contacts and abductees on a regular basis. Underground bases for their use would also be constructed, as well as bases for joint alien-human interaction.

By 1955 Cooper stated it had become obvious that the aliens had broken the treaty - as mutilated humans and animals were being found across the United States. It was also suspected that some abductees were not being returned and that the contact lists were incomplete. It was also discovered that the aliens were using humans and animals for a source of glandular secretions, enzymes, hormonal secretions, blood plazma and possibly genetic experiments. The aliens explained that these actions were necessary for their survival. They stated that their genetic structure had deteriorated and that they were no longer able to reproduce, and that their race would soon cease to exist if the genetic structure could not be improved. This explanation was looked on with suspicion.

However, since the weapons of the United States, whilst being easily the most powerful of any nation on

Earth, were literally useless against the aliens, it was decided that friendly diplomatic relations must be continued - at least until such time as the United States was able to develop technology which would be able to challenge the aliens militarily. So contact was maintained and, over the years of contact with these aliens, Cooper said that the Government has come into possession of technology beyond our wildest dreams, which includes craft which can now travel between planets. Whilst this technology was probably originally developed with good intentions for the defence of humanity, the rulers are now using the technology to control humanity itself.

Cooper was one of the first "insiders" to reveal the existence of extraterrestrials, but in the subsequent "Disclosure Project" created by Dr Steven Greer, over 500 government, military and intelligence community witnesses have stepped forwards to testify to their direct, personal, first-hand experience with UFO's, ET's, ET technology and the cover up that keeps this information secret.

The events described by Cooper from the 1940's onwards were largely dismissed at the time by the public, and government announcements dismissing the idea of crashed alien crafts were largely accepted. However, as disillusionment surfaced due to continuing lies, unethical conduct and morally debatable actions by government officials and agencies, more and more members of the public took the view that perhaps governments around the world were hiding something. Too many sightings of these unusual crafts and alien abduction activity were being reported, making the official debunking explanation harder and harder to accept. Many people were asking questions which were not being adequately answered.

The 2005 Foreword written by Budd Hopkins in the new 2005 edition of *Left at East Gate* by Larry Warren and Peter Robbins included the following conclusion on the subject of UFO activity:

"Many naturally looked to the government and the armed forces for explanations and reassurance, but as *Left at East Gate* amply demonstrates, the official policy seems to be to silence witnesses, to explain nothing, and to deny everything."

This view is not an isolated opinion. Linda Moulton Howe said in *Glimpses of Other Realities* that she received a document in a plain manila envelope in August 1995 which contained several pages and a title sheet that said "Restricted SOM 1-01 Majestic-12 Group Special Operations Manual" and was top secret. The complete document was reproduced by her in Appendix 1 of Volume 2 of her book (*High Strangeness*) and contains (amongst other incredible information) a statement that:

"*The official government policy is that such creatures [entities known to be of extraterrestrial origin] do not exist, and that no agency of the federal government is now engaged in any study of extraterrestrials or their artifacts. Any deviation from this stated policy is absolutely forbidden*".

In 2006, Jose Escamilla released a film entitled "UFO: The Greatest Story Ever Denied", which was an in-depth analysis of the UFO phenomena and was inspired by his own experiences at Midway, New Mexico where he personally encountered a UFO. The film contains incredible footage of UFO's, mostly filmed through infrared lens as UFO's are mostly not visible to the naked eye.

Are UFO's real? The truth is that even the US Military has never denied their existence. Indeed, John F Kennedy, while President of the United States of America, stated that "The US Air Force assures me that UFO's pose no threat to National Security". At the beginning of Escamilla's film he also quotes from a Report of the USAF Scientific Advisory Board issued in March 1966 when it was stated, "The Committee concluded that, in the 19 years since the first UFO was sighted, there has been no evidence that the Unidentified Flying Objects are a threat to our National Security".

Note that they said that UFO's *pose no threat*. They never said that *they do not exist*.

Moreover, as reported by Brad and Sherry Steiger in their book *Real Aliens, Space beings and Creatures from Other Worlds*, as long ago as 1968 UFO sightings had become so common that military radar screens them out with UFO filters, and that at that time the only system doing any recording of UFO's was the US space tracking system. In an interview the Steigers conducted with a NASA atmospheric physicist, it was reported that there were so many UFO's in the sky that the Air Force has had to specially screen them out.

The information provided by the Steigers' in their 2011 book is wide ranging and of great interest to those who do not possess their encyclopedic knowledge of the subject. A fascinating piece of information revealed by them in the book related to a top secret revelation made by Bob Dean, a retired Army Command sergeant major, during an interview with them during the early 1990's. Dean had been stationed in 1963 at the Supreme

Headquarters, Allied Powers, Europe (SHAPE) in Paris as an intelligence analyst and had access to a report commissioned in-house to assess the situation involving incidents relating to the circular discs and related matter. The report was first known as "The Assessment" and was later named "The Assessment: An Evaluation of a Possible Military Threat to Allied Forces in Europe". Dean felt that this information needed to be revealed, and on retirement he became one of the best known speakers on the UFO circuit. It caused him to develop an obsession with studying and researching history, philosophy, religion, anthropology, archaeology – essentially anything which might fill in the pieces and add to his understanding of the evolutionary, and other, implications of UFO's and extraterrestrials on our planet. Many have followed Robert Dean's pioneering path to seek knowledge of extraterrestrial existence.

So what did this top secret report published in 1964 reveal? The alleged findings are incredible. They include: (i) Our planet and the human race have been under in-depth surveillance for thousands of years. (ii) Because of the high level of demonstrated technology, whoever or whatever is behind the circular metallic discs could not be a real danger. If they were hostile or malevolent, they could have easily destroyed the human race. (iii) We have virtually no defence against their technology so it can only be concluded that they must simply be watching and observing. (iv) An alien plan does appear to be under development for eventual interaction with Earth's inhabitants.

The report then allegedly stated that there appeared to be four alien races visiting Earth: (i) The "Grays" (biological androids or clones); (ii) Humanoids or human-like in appearance; (iii) Taller Grays, about six feet tall but without the big wrap-around eyes; and (iv) Reptilians, with lizard-like skin and eyes with vertical pupils.

Are they really all non-hostile, or should we be worrying? The reports on alien interaction are varying as further reviews in the chapters on Contactees and Alien Abduction will reveal.

Global "Roswells"

The Roswell incident of July 1947 is the central icon of ufology, and so if the authorities could persuade the public to perceive this incident as a fantasy, the thorny problem of "extraterrestrial existence" might go away. That has never happened.

Indeed, since the UFO crash at Roswell, there have been numerous similar incidents reported and documented all over the world. Three books reviewed below reveal three key UFO encounters that have received worldwide coverage and have caused difficulties for the national security services of the countries in question: the Roswell crash itself in New Mexico, USA, the "British Roswell" UFO landing incident, and the "Russian Roswell" crash. The British and Russian incidents are referred to as "Roswell-type" incidents simply because of the publicity generated by them - bordering on mystical to some members of the public.

Although many do not realise it, the Roswell incident changed our lives, as will be apparent from the review below.

The "original" Roswell incident

The US Government disinformation on the Roswell incident has been a continuous process since 1947, with a 'final' explanation, fifty years after the event, by the US Government in 1997, which has only helped to make the public believe that it is being misled on what really happened.

The book entitled *The Day after Roswell* is therefore a critical source of information concerning the existence of alien technology and alien life forms. It is another milestone in the revelation of extraterrestrial existence. Whilst there have been many criticisms of the book, and some errors have been noted, the author, Colonel Philip Corso, can be considered a reliable source. He was intimately involved with these issues, and was a top military officer who had access within the system all the way to presidential level, and provided advice and direction at the highest level on the so-called extraterrestrial problem.

The story told by Colonel Corso is fantastic, and warrants a full review below. It is about a real alien threat to humanity. Anyone who reads the book, written just before his death in 1998, is likely to consider it to be a life-changing book, as it forces us to think beyond our "every day" safe and comfortable life. If what Corso writes is to be believed, there is a lot more to this world than meets the average humble human eye.

Who was Colonel Corso?

Corso was no ordinary guy writing about Unidentified Aerial Phenomena, Unidentified Flying Objects, alien space travel, alien technology and alien beings. He was a distinguished military officer, a member of the White House National Security Council under President Eisenhower, and headed the world-changing Foreign

Technology Desk at the US Army's Research & Development Department (more about his role and the specific activities of this Department below).

When Corso arrived at the R&D division of the US Army's activities in 1961 he inherited the Army's most closely guarded secret: the "Roswell Files". These files contained the cache of debris and information a US Army team had pulled out of the wreckage of a 'flying disk' that had crashed outside the town of Roswell in the New Mexico desert in July 1947. These incredible files had languished unopened for fourteen years since 1947, as the US Army had felt that if the files were opened, Soviet spies in the CIA would pass the information in these files back to the Soviet Union. The information was simply too sensitive for that, so they were mothballed. When Corso arrived in R&D, his role was to pass the technology information contained in the files to trusted scientists and industrial organisations who could 'back engineer' the Roswell crash wreckage inherited by the Army, so that the technology back engineered could appear as sudden and startling progress by US industry, together with registration of the necessary patents for the protection of the technology.

What Corso revealed

After the crash, a covert group known as "MJ12" was hastily assembled in 1947 as the military tried to figure out what it was that had crashed, where it had come from, and what the intentions of the inhabitants of the craft were. MJ12 was an alleged secret committee of scientists, military leaders and government officials formed in July 1947 by executive order of the then

President Harry S. Truman. Its main purpose following its formation was to investigate the recovery of the crashed UFO craft at Roswell. A huge cover-up operation immediately commenced after the crash with two aims: firstly, to keep the most sensitive facts about alien technology and visitation away from America's enemies and, secondly, to gradually desensitise the general public toward a future time when the reality of alien visitation would become public knowledge. It was felt in top secret government circles, primarily the MJ12 group, that it would cause mass panic and the breakdown of societal control to reveal this information suddenly.

The end result of the whole exercise was that incredible new technology came into existence, such as lasers, integrated circuitry and microminiturisation of logic boards, fibre optic networks, supertenancity fibres, accelerated particle beam devices or directed energy weapons, molecular alignment metallic alloys, HAARP, portable atomic generators, irradiated food, 'third brain' guidance systems (the EBE headbands), electromagnetic propulsion systems, depleted uranium projectiles, night vision goggles and bullet proof vests. Indeed, most of the technology that we take for granted now in the world of computers, communication and travel amongst many other things came directly from the back engineering of the Roswell artifacts. The military under General Trudeau described this as the "alien harvest". Nothing was the same again after the Roswell crash. The world changed forever.

Corso's view of the threat

Corso believed that UFO's were certainly here among us. He stated that from the very beginning of our endeavours to put satellites in orbit the extraterrestrials have been surveying and then actively interfering with our launch vehicles, "buzzing" our launched vehicles, jamming radio transmissions, causing electrical problems with the US spacecraft systems, or causing mechanical malfunctions.

He stated that the first report on extraterrestrial contact had been submitted by Lieutenant General Nathan Twining, head of Air Material Command at Wright Field Air Base, where the debris was shipped.

Twining is important to UFO researchers as he participated in a number of highly secret meetings relating to security issues posed by the discovery of UFO's at the White House during Eisenhower's tenure there. He subsequently became an adviser to President Truman. He reported that the UFO discs were a military threat due to their evasive action when "contacted" and their extreme maneuverability beyond anything known at that time. He described them as "circular or elliptical in shape, flat on the bottom and domed on the top." President Truman was briefed by Twining on the strangeness of these 'extraterrestrial' crafts. They seemed to have no engines; no fuel; nor any apparent method of propulsion, yet outflew the fastest fighters the USA possessed; the ability to see daylight from inside even when it was dark outside; the metallic fabric which did not burn or melt; the thin beams of 'laser' light that emanated from the craft; and much more. Further, there were odd child-like creatures inside the crafts.

The "alien threat" was considered real enough that the "Cold War" was used as the excuse/reason for developing defence systems that looked into space as well as Earth: the so-called star wars program.

We are now in possession of alien technology to assist our protection from them, if Corso is to be believed. It appears that one reason that the Reagan Administration pushed so hard for the Space Defense Initiative (SDI) in 1981 was the alleged 122 photos taken by astronauts that showed evidence of alien presence on the moon.

The reality according to Corso when developing SDI was that humans were vulnerable to the UFO's that "soared around the edges of our atmosphere, swooping down at will to destroy our communications with EMP bursts, buzz our spacecraft, colonise the lunar surface, mutilate cattle for biological experiments, and even abduct humans for their medical tests and hybridization of the species". The entities entering our atmosphere appeared to be genetically engineered humanoid automatons, cloned biological entities, who were harvesting biological specimens on Earth for their own experimentation. Corso concluded that until we developed a defence system, we were essentially farm animals to be harvested, and always had been. This was clearly a very negative situation that humanity found itself in.

Corso's view was backed by General Douglas MacArthur, who after retirement said in 1955 to the New York Times "The nations of the world will have to unite, for the next war will be an interplanetary war." This was taken little notice of, but was in fact the key strategic thinking of the military at that time.

Corso's description of the 'Grey'

Corso says he witnessed what we now describe as an extraterrestrial "grey" in 1947, a preserved specimen from the crash, floating in preservation semi-liquid gel. The skull of the creature, according to Corso, was overgrown to the point where all its facial features were arranged frontally, no ears, cheeks or facial hair, with a slit for a mouth. He described the 'greys' as "designed for space travel", with flexible skin to enable them to cope with anti-gravity g-force movement, different digestive systems so they could survive for long periods without sustenance, night vision anti-glare eye covers, but that they were "very human" notwithstanding their different appearance.

From the autopsy reports the greys were described as being "about four feet tall, with organs, bones and skin composition different from humans, bigger hearts and lungs, with skin designed to protect the vital organs from cosmic rays, wave action or gravitational forces not yet known to us." Yet the report noted how similar they were to us. Corso considered that they were perhaps genetically engineered and designed to adapt to long space voyages within an electromagnetic wave environment at speeds which create the physical conditions described in Einstein's General Theory of Relativity.

Corso goes on to give a lot of detail about the greys and their craft. It seems that biological time must have passed very slowly for the grey entity because it possessed a very slow metabolism, evidenced, according to the Walter Reed examiners, by the enormous capacities of the huge heart and lungs. The physiology of the entity indicated that its body didn't need to work hard to sustain itself. The larger heart drove a thin, milky lymphatic-like fluid (which was not blood) through a limited, simpler, reduced-capacity circulatory system. The blood and lymphatic systems appeared to be combined. Thus, the biological clock beat slower than a human's, and would allow the entity to travel great distances in a shorter biological time than humans. Moreover, the heart worked less hard than the human heart muscle because the entities were meant to survive within a reduced gravity field. The lungs had the capacity to "store" atmosphere for long distance travel, as did the skeletal tissue, with more flexible bones, like shock absorbers. They were perhaps not meant to exit the craft as they had difficulty breathing in our atmosphere.

They had no food or waste product, so it appears that they had a digestive system similar to plants. They had a one-piece protective covering type of outer skin in which atoms were aligned so as to provide greater tensile strength and flexibility. This would have protected them against low energy cosmic rays that would routinely bombard any craft during a space journey. Without the "suit" the entities would have probably experienced conditions similar to being in a microwave oven. The particle bombardment would have "cooked" the creature. The "inner skin" had a thin layer of fatty tissue that was completely permeable, as if it was constantly exchanging chemicals back and forth with the combination blood/lymphatic system.

As an aside, it is said by others who have reported on the physical make-up of Greys (for instance Nigel Kerner in *Grey Aliens and the Harvesting of Souls*) that they appear to be assembled out of some kind of biological tissue or mulch, interspersed in an amalgam of mercury and the finest gold wires. No sexual organs or reproductive processes have apparently been found on Grey bodies at autopsy. It is said therefore that this is where human DNA becomes important in Grey reproduction and is one reason why abduction of humans may occur.

The details provided by Corso have also been verified by others. For instance, in his book *Need to Know*, Timothy Good references the book by Dr Leir, *UFO Crash in Brazil*, which also gives immense details about a similar creature after it was taken to a local hospital in Brazil following the crash of an alien craft that it was piloting.

Communication and the Craft

The system of communication involved no words, but what could be described as mental telepathy or more like empathic projection, as the communication also contained an emotional component, in this case sadness for their dead companions following the crash. The brains used "headbands" as sophisticated communication devices to contact their mothership it was believed, and possibly for navigational purposes too.

The craft itself operated differently from anything ever seen on Earth. The craft was able to displace gravity through the propagation of magnetic waves controlled by shifting the magnetic poles around the craft so as to control, or vector, not a propulsion system but the repulsion force of like charges (electromagnetic anti-gravity propulsion). Once the USAF realised this, they apparently raced to figure out how the craft could retain its electric capacity and how the pilots who navigated it could live within the energy field of a wave.

They discovered that the craft functioned like a giant capacitor (storage of energy). In other words the craft itself stored the energy necessary to propagate the magnetic wave that elevated it, allowed it to achieve escape velocity from Earth's gravity, and enabled it to achieve speeds of over 7,000 mph. The pilots weren't affected by the tremendous g-forces that build up in the acceleration of conventional aircraft because to aliens inside it was as if gravity was being folded around the outside of the wave that enveloped the craft, in other words, like travelling inside the eye of a hurricane. Their suits enabled them to literally become part of the electrical circuitry of the vehicle, becoming an extension of their own body.

Essentially the entities were able to survive extended periods living inside a high-energy wave by becoming the primary circuit in the control of the wave. The alien technology was so far beyond Earth-based technology at the end of the Second World War that it is possible that even today the full extent of the "alien" technology may not be understood by us. Right in front of the humans who witnessed it, however, was the formula for long distance space travel.

Dr Herman Oberth suggested we consider the Roswell craft not as a spacecraft but as a time machine. Were these beings perhaps humanoid robots rather than lifeforms, specifically engineered for long distance travel through space or time?

Verification

Paul Hellyer, the former Canadian Defence Minister, has gone on record as saying what Corso put in his book is true from his own investigations, and he maintains that the US has now developed the aliens' own weapons to the point where they can be used against the aliens.

It indeed may well be that the US and other nations have developed craft that can travel in space. Circumstantial evidence produced by researchers such as Timothy Good suggests that this is the case. Whether or not we can now travel to 'them' in the same way they can travel to 'us', even top military men state that there is too much evidence to deny that something unexplained is going on in Earth's atmosphere.

The "Code Orange" story

In his book entitled *Earth: An Alien Enterprise*, published in 2013, Timothy Good writes about an array of fascinating topics across the extraterrestrial spectrum, including information about why aliens have been visiting Earth, detailed stories of contacts between members of the public and aliens, UFO evidence, technology developments through alien contact and much more. However one of the most fascinating stories that does not seem to have been repeated in other literature is the story of two extraterrestrial biological entities who arrived when the Roswell craft(s) crashed, and came to England over the period 1955-57 for rest and recuperation as they were unwell at their base in the USA. The exercise was known as "Code Orange" according to Good in the chapter entitled "Gray Liaison".

It was the incredible story of a period of time during which six RAF airman guarded the aliens and made sure they were happy and that their needs were met. The chapter tells of the discussions that ensued between the airmen and the aliens, during which they each grew to care greatly about each other. The aliens warned of matters such as the future overpopulation of Earth, and the poisoning of the environment which would have catastrophic consequences during the 21ˢᵗ century, but of interest here is the explanation by the aliens of why the Roswell incident had occurred. The simple explanation given in the section entitled "Reasons for Roswell" was that the aliens wished to offer the US and British military and governments "*scientific instruments to forward human progress rapidly and without cost*". The story of Code Orange is reason enough to acquire Timothy Good's latest book, but the information throughout the book is well presented and highly readable.

The British "Roswell" incident

Left at East Gate, the 1997 book by Larry Warren and Peter Robbins concerning the UFO landing at Rendlesham Forest in East Anglia in December 1980, is an incredible book for one reason: It is a *personal* experience. There are many books detailing UFO incidents, but most of these are the product of hard working UFO researchers, not someone's personal recounting of an experience that can only be described as "out of this world". Larry Warren's story, and that of Peter Robbins, who joined him to write the book, are both life-altering, and it would be a brave person to want to go through the harrowing experience both men recounted.

The story centres around Warren's recollection of the landing of a triangular-shaped craft of unknown origin which illuminated the entire forest with a white light, beams of light shooting down from above, alien entities suspended in mid-air beside the craft seemingly repairing the craft, interaction with the alien entities and US Air

Force officers, background radiation readings, damage to trees near the landing site, and marks on the ground at the site.

The events were clearly shocking, not least because the landing occurred on the fencing outside two air force bases, RAF Bentwaters and RAF Woodbridge, that secretly housed the largest stockpile of tactical battlefield nuclear weapons in the whole of the NATO infrastructure.

On the morning of 29 December 1980, after the third night of UFO encounters, specialist officers are said to have arrived to check damage to the nuclear arsenal, as the UFO's had fired pencil-thin beams into the bunkers housing the nuclear weapons. It was, and still remains, the desire of the UK and USA Governments desire to keep the events of those fateful nights secret. As Larry Warren and Peter Robbins indicate, the reasons why the "security lid" was slammed down so hard are entirely justifiable. The idea of unidentified flying objects being able to freely violate the airspace over the nuclear weapons storage area and fire laser-like beams of light into the compounds hardened bunkers is unacceptable.

Warren, a member of the RAF Bentwaters security police at the time of the incident, recounted some incredible events. He stated that during the encounter involving the alien entities, they interacted with the base commander, who was a tall man, and actually cocked their heads back slowly so they could see his face better. At this point he realised that these entities were really alive. He believed there was telepathic communication.

The morning after the encounter, the personnel involved were debriefed in what Warren recalled as an underground facility during which the debriefing officer stated that numerous civilisations visit Planet Earth from time to time, and some have a "permanent presence here". He also believes he telepathically heard the voice of an alien who stated that its race had blended into earth's society at all levels, and the underground facility that Warren was in at that time was one of many processing zones throughout the world. Warren surmised that UFO crafts entered the underground facility through an extensive tunnel system. During his research, Robbins did manage to find a file containing security plans for major subterranean construction under Bentwaters between the years 1966 and 1968, which gives credence to Warren's claims.

Warren believes his mind was "messed with" to remove specific memories of the event. He was told to sign a statement that he had only seen some unusual lights in the trees, nothing more. As an aside, Linda Moulton Howe reported in Volume 2 of her book *Glimpses of Other Realities* that she had interviewed a witness from a 1971 encounter in the Cambodian jungle where the witness (an experienced member of US Special Forces) had encountered extraterrestrial humanoids. Her book is full of incredible interviews, but the interesting part of this interview was that after the event the witness underwent drug and hypnosis treatment to alter his memories of the incident. Further, she interviewed a member of security at RAF Bentwaters, Staff Sergeant James Penniston, who backed up Warren's belief that narco-hypnosis was used to alter his memories of the incident at Bentwaters. This process of memory erasure after serious extraterrestrial encounters would appear to be part of the required government cover-up referenced in earlier sections.

The landing of the craft received worldwide coverage on 1 October, 1983 when the mass-circulation British newspaper *The News of the World* broke the story with a headline that read "UFO lands in Suffolk: And That's Official". The paper quoted, in particular, a document issued by the Deputy Base Commander on 13 January 1981 which, amongst other things, confirmed the appearance of a metallic object which was triangular in shape and approximately two to three meters across the base and two meters high. The object had a pulsing red light on top and a bank(s) of blue lights underneath. The object was hovering or on legs. The article also contained a drawing of the craft done by Larry Warren. This was followed by a further article on 6 November 1983 with a more lurid title "Bug-Eyed Alien Greets Air Chief" providing further information including details of the description of the aliens that manned the craft.

Warren and Robbins revisited the site in 1988 and believe they had a further encounter with aliens and a craft. Many pages are devoted to transcripts of the night when the two men encountered a UFO and alien presence. Whatever the occurrences of that night on 18 February 1988, both are convinced that these life forms are real, have their own agenda and come and go with impunity. Warren believes he has had multiple unwilling encounters with aliens through the years, and even underwent regressive hypnosis with Bud Hopkins.

Perhaps the most unnerving side of the story of Larry Warren and Peter Robbins is the national security aspect. From their experiences, they believed that their country was slowly becoming a national security state. Robbins described the USA as "A place where keeping secrets had proven to be more important than safeguarding the most decent aspects of our American democracy". This secrecy was about hiding alien visitation as much as any other threat.

At one point Robbins stated in a tape recording: "We are here to try and shed a little more light on why there's so much terror in the truth." That statement is a good summary of the essence behind the remarkable, nearly 500-page long, book by Warren and Robbins entitled *Left at East Gate*.

Nick Pope, the former UFO spokesperson at the UK Ministry of Defence, mentioned in *Open Skies, Closed Minds*, that the most interesting cases relating to UFO activity were near to UK military bases, most notably at RAF Cosford, RAF Shawbury, and of course RAF Bentwaters. These sightings were observed by credible servicemen and at this time all these cases remain with an unsatisfactory conclusion.

Could it be that perhaps the aliens wish to observe top secret airforce base activity?

A note about Wilhelm Reich

An interesting side issue which arose out of Warren and Robbins' book was the references to having seen a "cloudbuster" at the air force base. A "cloudbuster" had been invented by Dr Wilhem Reich, and was a rather unusual-looking device that could allegedly alter weather and was also said to be able to attract UFO's. Peter Robbins mentioned this to Larry Warren, and said he had seen a demonstration of one in operation, and they actually worked. When shown a picture of one by Robbins, Warren then recollected that he had seen a scaled-up version of one at the air force base.

Warren's recollection was confirmed by another member of staff at the base who was interviewed in August 1991.

A cloudbuster creates atmosphere movement by drawing energy down from the atmosphere, then grounding it harmlessly in moving or deep water. Cloudbusters had been proven to work, and Robbins surmised that perhaps a scaled-up one could have caused the "freak storm" that had flattened the forest on 16 October 1987. On 9 March 1994, the UK's Forestry Minister, Earl Howe, planted an oak tree at the site to mark the two millionth tree planted in replacement of the one million trees blown down in the storm.

As a brief aside, Reich considered in the 1950's that he had discovered the healing life force of the universe, which he called "orgone energy", or "life energy", and which he attempted to capture for healing purposes in his "orgone energy accumulators" or "orgone boxes", as they were sometimes called. Patients sat inside the boxes to benefit from the healing energies that were reputed to exist, which allegedly included a feeling of well-being, healing body wounds and, some said, cancer healing through building up the body's immune system. The US Food and Drug Administration took a dim view of the sale of these boxes, which some might consider to be the sale of "earth energies", and in 1956 stopped their sale. Reich was sentenced to two years imprisonment for breaking the court order, and when his appeal against the sentence was rejected during 1957, he was sent to jail, where he died that same year.

Russia's "Roswell" incident

A brief review is included below of two interesting incidents from Russia which are included as a result of the 2012 publication *of Russia's Roswell Incident* by Paul Stonehill and Philip Mantle. The book also contains details of a huge number of other incidents in the Former Soviet Union, including UFO sightings at secret Soviet military bases such as Kapustin Yar. Russia is subject to UFO observation at new technology test sites in the same way as has occurred in the USA and elsewhere.

The two world famous "UFO" incidents reviewed below were at Dalnegorsk and at Tunguska.

The Dalnegorsk incident is an intriguing crash on 29 January 1986 of a UFO into a hill near Dalnegorsk. This crash came to the attention of the American public in 2012 as glassy spheres and bits of metal from the crash were put on display at the National Atomic Testing Museum in Las Vegas. The sphere that crashed eventually took off again, but left behind 'fine mesh', and 'small spherical objects' amongst other things.

A CIA brief on the crash included the following statement: "Some of the scientists have concluded that the object that crashed into Hill 611 was an 'extraterrestrial' space vehicle constructed by highly intelligent beings. Doctor of Chemical Sciences V Vysotskiy stated that "without doubt, this is evidence of a high technology, and is not anything of a natural or terrestrial origin". He went on to cite evidence of why the properties discovered were not from earth.

The display case at the Las Vegas museum stated "Three Soviet academic centres and eleven research institutes analysed the objects from this UFO crash. The distance between the atoms is different from ordinary iron. Radar cannot be reflected from the material. Elements in the material may disappear and new ones appear after heating. One piece disappeared completely in front of four witnesses. The core of the substance is composed of a substance with anti-gravitational properties".

Apparently elements of gold, silver and nickel disappeared when melted in a vacuum and were replaced by molybdenum, which was not present at the start of the experiment and does not occur as a free metal on earth.

The book also details the 1908 Tunguska crash which has also posed a mystery ever since the event occurred. The authors state that "something" burst in the air over Siberia flattening more than 1,300 square miles of forest but leaving no crater and no evidence of what caused the event. Some investigators have claimed that it was an unidentified flying object, perhaps a mother ship, that exploded possibly due to a malfunction of its onboard power plant. The force of the explosion was considered to be 2,000 times the force of the atomic bomb that exploded over Hiroshima in 1945.

In the aftermath of the Tunguska explosion, a strange illumination lit up the night and lasted 72 hours - so bright that the glow was sufficient to light the streets of London during the night time hours. What really did happen?

Project Serpo – travelling to the stars

The fears expressed by Colonel Philip Corso in *The Day after Roswell* related to possible alien invasion of Earth. Was there really a threat to the human population of invasion or the takeover of Earth? Len Kasten's *Secret Journey to Planet Serpo* (2013) makes it clear that this threat was indeed real, just as Corso had stated.

As we will see later, "Project Serpo" was a tale of travel to a planet 39 light years from Earth, where twelve fearless and courageous humans lived side-by-side with an alien race for thirteen years. Len Kasten's book is a classic turning point in our understanding of extraterrestrial existence, and no summary can do it justice. The book simply demands to be read by anyone who has an interest in this hidden extraterrestrial world.

The conclusion can only be that *they exist*; that is the reality. The information presented by Len Kasten, one of the world's foremost researchers in this field, is so important that the essence of his writing is contained below.

To understand how the incredible reality came about that humans travelled to a planet that is 39 light years away, or 240 trillion miles away, we need to go back in time to the end of the Second World War when the threat of alien invasion and takeover of the planet was a genuine fear of the US military, as Corso and William Cooper indicated (see previous chapters).

The USA, as the most powerful nation on Earth after the War, was just settling down to peace time again when its military leaders became aware of a much more serious threat than those posed by terrestrial attack: the possibility of extraterrestrial alien attack utilising weapons way beyond our technological capability at that time.

The craft that crashed at Roswell was not the first encounter the US military had experienced with antigravity crafts which could out-maneuver their fastest and most sophisticated aircraft. However, the Roswell crash raised military fears because it appeared that the craft that crashed had been spying on sensitive military installations, particularly those with a nuclear application, and not least the Roswell Army Air Field at Roswell which was the base of the B-29 squadron that had dropped the atomic bombs on two Japanese cities a mere couple of years previously.

This type of surveillance raised suspicions because, with the information the US military already possessed, it was not a difficult conclusion to reach that aliens might have been seriously planning an invasion of the planet. Any resistance to an invasion would undoubtedly be most likely to occur from America, so it would be natural for aliens to be surveilling current US defence and attack capability, and working out what weaponry to negate.

As mentioned earlier, the US military knew about antigravity crafts and their superior capability before the 1947 Roswell crash. How did they know this?

Len Kasten's book takes us back to the Nazis and their base in Antarctica during the 1930's and 1940's where it was discovered that they had been developing antigravity craft that could hover, fly at supersonic speeds and at high altitudes, and change directions instantly. These crafts were superior flying machines and effectively left the Allies defenceless against attack from such crafts.

A galactic war on Earth

Fantastic though it may sound, it is suggested that Earth may have been used by two warring groups of extraterrestrials as the venue to fight their war against each other. Hitler had teamed up with an alien race which was transferring advanced technology to him. This race of reptilian, lizardlike extraterrestrial beings came from Alpha Draconis and Orion and formed the Draco-Orion/Grey Empire and had their base in a massive underground complex that stretched from Tibet to India. These are said to be the mythical serpent people of Hindu mythology, who were cruel and merciless enslavers and masters of mind control. These extraterrestrials had been working with the Nazis to transfer specific advanced technology and perfect a distinctive bell-shaped craft utilising antigravity or electromagnetic propulsion technology unknown on Earth at that time. This craft was called the "Haunebu" which had been manufactured and tested in various increasingly advanced versions for use in battle. Had the War lasted much longer, it appears that the Nazis would have had complete air superiority through the use of this craft.

Lined up against them were their ancient enemies from the humanoid races of Andromeda, Arcturus, Lyra, the Pleiades, and Sirius. This group is sometimes called the Galactic Federation of Light. The so-called Federation races are non-interventionist, believe in freedom and free-will choices and seek to assist in spiritual development. They are said to have bases on Earth under Death Valley and Mount Shasta. They supported the Allies through working with Nikola Tesla, which allowed the Allies to develop radar to counter the advanced technology introduced by the reptilian forces linked with Hitler. It seems that in the end the combined brainpower of Roosevelt and Churchill and the resourcefulness of the thinking and alert Allied soldiers won the war. Whether true or not, the idea that supporters of the concept of free will rather than slavery won through is a comforting thought.

Neuschwabenland - Antarctica

The location of the reptilian Hitler-supporting base was the reason why the Nazis possessed a large territory in Antarctica which they had named 'Neuschwabenland' and had established a large city under the ice called 'Neu Berlin' where around forty thousand scientists and others lived. As will be seen later ("They left signs that they were here"), this land may well have been the 'lost' land called Atlantis. It could have been devised as a refuge in the event that the War turned out badly but, as Kasten points out, it is more likely that the colony established was a base for joint German-alien scientific and technological development for interplanetary travel and conquest. Because of this pact with the reptilians, the Nazis would have been secretly supremely confident of winning the war.

Len Kasten references information from a source known as "Branton" that the Draco Reptilians, in alliance with extraterrestrials from Orion, have already conquered and enslaved civilisations in twenty-one star systems in a nearby section of the galaxy. Hitler's association with such powerful allies, may well have been the reason that he considered himself invincible.

Moreover, the colony was intended to preserve and to create a master race. It was reserved for only the purest specimens of the Aryan race, a place from where Hitler's 'master race' could be safely bred. It has been reported that Neuschwabenland had been placed under the control of Heinrich Himmler by mid-1943, and he had selected ten thousand of the "racially most pure" Ukrainian women out of the half-million ethnic German women who were deported from Russia, and sent them to Neu Berlin on the submarines. They were all blond blue-eyed and between the ages of seventeen and twenty four. He also reportedly sent 2,500 battle-hardened Waffen-SS soldiers who had been fighting on the Russian front. At four women to each man, they were expected to breed the Aryan population of the new civilisation under the Antarctic ice. Hitler himself may even have ended up there, with his 'double' executed in the Berlin bunker rather than him, as the Soviet leader Joseph Stalin is reported to have said "Hitler escaped and no traces of him were found."

This colony, with its superior weaponry and extraterrestrial friends, was intended to be in a position to take over the world in due course and enslave the "inferior" races on the rest of the planet. Through

British intelligence, the Allies came to know of this colony, and allegedly attempted to break it up in the period from the end of the war until early 1947. However, these operations seem to have been a failure, and the antigravity discs and other alien-derived weaponry utilised by the Nazis and their alien friends may have been the reason for this failure. This scenario had also been written about by Mattern Friedrich in his 1975 book, *UFO's: Nazi Secret Weapon?* Certainly Admiral Byrd made a grave warning to US Naval intelligence in 1947 about what he had discovered in Antarctica, although he did not specify what that threat was.

So in 1947 the most powerful nation in the world was debating in its highest circles how to stop this new flying technology that was being developed in Antarctica and that was threatening to undo the world peace that America had just won at such cost, when an event-changer occurred: the Roswell craft crashed and a solution presented itself.

Serpo

This is where the Serpo story, the most amazing story of benevolent alien contact, begins.

We know much of the information relating to "Project Serpo" because a former official of the Defense Intelligence Agency (DIA) who had been assigned to this top-secret government project started releasing emails in 2005 onto a website which was set up specifically to receive his revelations. The releases started on November 2, 2005. This was exactly twenty-five years after the final report on the project (then known as Project Crystal Knight) was written. This was the earliest date that the report could be declassified under government policy rules relating to disclosure of secret documents and therefore legally disclosed to interested members of the public. It is not believed that this was a rogue operator who released this information, but more likely was authorised for release at the highest level. This is because of the way the DIA had been set up by President Kennedy, who had wanted to create an agency that was transparent towards the public.

It is reported that two ET spacecraft crashed in July 1947 and the US military was able to collect all the debris, including one alien that had survived the crash. All others died. The alien (a male) lived until 1952 and was named "ebe1" as he was the first "extraterrestrial biological entity" that it had been possible to communicate with. It transpired that this particular alien was friendly and was able to assist the USA in its desire to receive the transfer of advanced technology that could match or exceed that received by the Nazis. There was a new and more serious war out there that the world knew nothing about, involving technology that the world's most powerful nation needed to understand and to develop if it was to survive in any impending interplanetary war.

Over the time that ebe1 was alive, it was discovered that he emanated from the Zeta Reticuli star system, and his race had developed, amongst many other things, a means of swift travel from one star system in the galaxy to another utilising 'space tunnels' which they knew how to locate. These are also commonly known as 'wormholes'.

He was able to communicate with his home planet utilising a communication device in his craft, and it was through this communication device, which US officials learned how to operate, that the idea of an exchange program with their planet was mooted. Communication with the "ebens" (as they were named) was initially difficult as the ebens didn't speak English, and spoke with a strange tonal language. The exchange program finally became a reality, however, in 1965. This venture was encouraged by the military to assist technology transfer, and reports suggest this has now been successfully achieved and we have the ability to do many things which have not yet been revealed to the world at large. We can certainly defend ourselves against alien attack in a way not possible in 1947.

Twelve US military men were trained for the mission, and the "eben" craft landed on 16 July 1965 to collect the humans for the trip. The landing and boarding of the alien craft can be visualised, as the film "*Close Encounters of the Third Kind*" (1977) depicted the events almost precisely. The trip itself can also be visualised as the subsequent film "*Contact*" (1997) depicted the journey and arrival on the foreign planet.

Len Kasten's book gives remarkable detail about the journey, the stay on Planet Serpo (which lasted thirteen years) and the return to Earth. The trip was logged in diaries maintained by the crew leader, and on no less than 5,419 cassette tapes that contain voice recordings of all aspects of their existence on the planet, ranging from the journey, living conditions, eating, exploration of the planet, death, technology, cloning techniques, star wars with other alien groups and much more. By the end of the book you simply cannot say that the story was invented. It would be impossible to have faked, for instance, 5,419 tapes lasting 90 minutes each which were carried with them to Planet Serpo. It took seven years alone to transcribe them all after their return.

"Eben" technologies that we allegedly got access to included free energy devices; particle beam weapons; antigravity land vehicles; cloning and creation of artificial life-forms; the ability to view/record past events (known as the "yellow book"); interstellar communications; translation devices; and sound weapons. The details revealed to the Serpo website (through the work of the website's developer Bill Ryan) included the fact that there is a book called "The Red Book" which has been updated every five years since 1947 and is available only to top government officials involved in secret UFO investigations and contact with extraterrestrials. It contains the most compelling cases, including analysis of trends, types of sightings, human contact with ET's, and national security concerns. The US President receives a summary at the time of each update.

The Serpo website noted that there have been visitors from nine other star systems. These included (amongst others) aliens from a planet near Alpha Centauri A ("the Greys"), and visitors from a G2 star system in Leo and Epsilon Eridani. Some aliens have been benevolent and some hostile. Some were hybrid beings created in a laboratory and some were created by natural birth.

In the end the book leaves you with no doubt that the authorities are well aware that we are not alone. *That they exist.*

PART 4

Contact (positive and negative)

Contactees – The Positive, Knowledge-Enhancing, Experiences

Timothy Good's 1998 book *Alien Base*, (subtitled *The evidence of extraterrestrial colonization of Earth*) detailed vast amounts of information received from 'contactees' and 'abductees'.

'Contactees' are people who have been "contacted" by human-looking extraterrestrials and who have had good experiences from the 'contact'. They have often received important messages, often anti-war messages or warnings about the destruction of our environment, or the extraterrestrial's worry that the failure of planet Earth (including the use of nuclear weapons) could upset the balance of other planets in our, or a nearby, planetary system.

'Contactees' are people with very different experiences from 'Abductees' (which are the subject of the next chapter). 'Abductees' are generally people who have had frightening experiences, often involving aliens who, though humanoid, are somewhat 'otherworldly' in physique and behaviour. The abductees often can only recall these experiences under hypnosis as their memory of the incidents has been blocked deliberately, although they do also receive messages.

Reports of 'abductee' experiences tend to involve invasive procedures on board alien crafts that are contrary to the free will of the individual. Such activities lead to a conclusion that not all aliens have a benevolent agenda where humans are concerned. This is in sharp contrast to the experiences of 'contactees', who have had knowledge-enhancing, and spiritually-enlightening experiences, which are detailed below.

Ridicule

It is worth noting that many 'contactees' as well as 'abductees' do not report their contacts for fear of ridicule, particularly by the media. The implication that someone has been talking to "little green men" is good for a cheap and easy laugh almost any time. Whether such reporting is due to fear or ignorance, it is unhelpful.

The story of Daniel Fry is a case in point. During his alien contact experience, he was asked to reveal his experiences to the press. He said he would be ridiculed, and the human-like alien replied: "Of course you will be ridiculed. Ridicule is the protective barrier which the timid or the ignorant erect between themselves and any possibility which frightens them...It is the price exacted from every individual who is as much as one step in advance of his fellows."

The contactee revelations

Timothy Good's reporting of the stories of 'Contactee' experiences do, however, make riveting reading and, apart from anything else, leave one with a warm and happy feeling. There are nearly 600 pages of astounding storytelling, including the tales of George Adamski, Daniel Fry, Albert Coe, Sir Peter Horsley, Howard Menger, Carroll Wayne Watts, Ludwig F Pallmann, Orlando Ferraudi, Sidney Padrick, Madam X, and many others.

Of the wealth of interesting information imparted by the contactees, detailed below are some of the more fascinating pieces of information.

It should be remembered, however, that 'proving' their experiences is futile. It is simply not possible. It was merely considered by most contactees to be a rare privilege.

Telepathy: Daniel Fry and Albert Coe both reported the use of 'mental telepathy', or 'extra-sensory perception'. The human-looking alien visitor stated that all humans can read the thoughts of other minds, but at the moment the humans on Earth are at an early stage of development. At this point in human development, the body's normal perception equipment is not being utilised because telepathy is a rather public form of communication. Currently, the individual requires a considerable degree of privacy in his words and thoughts. Earth's animals use this sense to a greater degree than humans, and for some of Earth's insects it is the only form of communication. Once humans understand better the need for transparency in their dealings, mental telepathy will change society.

Menger also stated that once he realised that the visitors were able to read his every thought, and he suddenly realised that it is not possible to hide anything, he became completely honest, not just with the visitors, but with himself.

The make-up of Earth's inhabitants: Menger said that although the space visitors were far superior to Earth humans in terms of physical, mental and spiritual abilities, they were still much like us. Ferraudi went on to say in his encounters that similarity to the visitors was due to our origins, as the human inhabitants of Earth were made up of five races, none of whom originally came from Earth. They are only remnants of civilisations that came from other planets.

In her book *No More Secrets, No More Lies*, Patricia Cori also reported this 'seeding' of Earth, though this was through channelled messages rather than contactee activity. She said however that there were four distinct seed races engineered, utilizing the blueprint/grounding of the genetic material of the indigenous earth-beings (the species called 'homo sapiens'). Formulating the master genetic coding was a collective effort of the donor extraterrestrial races, and light beings from many realms participated in the process. The selected primary races were considered prototypical of the elements of Earth, and they held resonance with four primary colors: black, red, white and yellow. They were placed in climatic conditions that most closely resembled their home planet climate. Whatever the process, this is yet another person propounding the theory that the current version of humans was genetically engineered by alien races.

As was discovered during the Project Serpo experience, which is referenced earlier, the fact that the genetic make-up of different beings or entities may appear different does not necessarily mean that the entities cannot live together in the same environment. Whether entities that are differently seeded can live together depends on their conduct, and whether the races are benevolent or hostile. The 'Ebens' which were referenced in the Project Serpo experience were benevolent, although of a reptilian appearance. The Ebens stated that other races around the star systems were not always as benevolent, and "star wars" had been fought with aggressive and hostile entities in the past.

Patricia Cori makes the interesting claim that following the completion of this genetic engineering exercise, the Annunaki (the Controllers) then deactivated ten of the twelve strands of DNA so that humans could be no more than slaves, changing the carefully laid plans of other alien races for the earth-based humans, and surrounded earth with an electromagnetic grid to block contact from the many light beings and planets that provided the seeding. Perhaps this was done in addition to the more obvious methods of control outlined by William Bramley, which is summarised in the later chapter "Global Control of Humanity". She says many humans are now reconnecting to their full 12-stranded DNA and awakening to their true purpose as light beings as they break free from the Annunaki hybrids that rule Earth. Many researchers have recently stated that the current control over humans on Earth is ending and a new era of freedom and transparency is taking the place of the current control matrix.

But returning to the subject of genetic engineering or manipulation, in his book *Cosmic Voyage*, Courtney Brown used the technique of remote viewing, and was able to ascertain much information about the Greys. This race of extraterrestrials have long been associated with human abduction for genetic rehabilitation of their race, and Courtney Brown's view was that the genetic manipulation work on humans was to incorporate from humans lost aspects of their own genetic make-up. He stated that the Greys are not satisfied with the wonders of their technology, their ability to travel effortlessly through time and space, their telepathic abilities. What they really want is one piece of our genetic make-up: the ability to *feel*, to have emotions.

A universe teeming with intelligence: Dan Sherman told an extraordinary story of his training to be what he called an "intuitive communicator" in an autobiographical book entitled *Above Black*. This is not "channelling" alien information like Patricia Cori, where you essentially "take on the identity" of the person you are channelling and they largely write the story for you. He was simply receiving alien-derived data which he passed on to the US military, though he did learn a little of the existence of the alien race he was communicating with. His story begins when he was informed while in the USAF that he had special human skills as a result of genetic management by an alien race while he was in his mother's womb, which allowed him to communicate with an alien race. He never met the aliens; he just communicated with them after being trained by the US military in how to activate and utilise his unique transcribing skill. In the end, after several years of passing on alien coded data, he grew tired of not knowing the purpose of his "downloads", and the extreme isolation his work caused to his personal life. His clear conclusion though (from his story as a participant in the US military project entitled "Project Preserve Destiny") was that alien existence was real. He also stated that alien contact projects are hidden from view behind other secret projects within US Government circles so that they remain away from political scrutiny. Over time he asked many questions of his alien contacts, but one answer from his contacts was that there is a vast number of other "intelligences" in the Universe. This was reiterated in the stories told by Daniel Fry and Air Marshall Sir Peter Horsley.

Fry and Sir Peter both reiterated that during their contact with the human aliens they were told that we live in a galaxy that teems with life and intelligence. Sir Peter's experience was one of the more fascinating tales, which he recounted in his extremely readable autobiographical book *Sounds From Another Room*. He was introduced in 1954 through reputable sources to a mysterious "Mr Janus" in a London flat owned by a 'Mrs Markham'. Mr Janus appeared to be able to read his mind. Although technology had not advanced anywhere near the point it has today, Mr Janus talked about the coming striving by Man to break his earthly bonds and travel to the Moon and beyond. This has now happened.

Mr Janus went on to say that man in his journeys through the universe may find innumerable centres of culture far more ancient than his own, an infinite variety of forces as yet unknown, even other universes with different space and time formulae. He said Man is preoccupied with his technological toys, but they will not bring him happiness. His culture is superficial, with a careless disregard for nature. On the question of when the aliens would be coming here, Mr Janus said that the alien traffic to Earth is just a thin trickle on the vast highways of the universe. Earth was a galactic backwater inhabited by only half-civilised men, dangerous even to their neighbours. He went on to state that Earth is being monitored mainly by robot-controlled space probes (note: probably like the craft that crashed at Roswell), though some are manned in order to oversee the whole programme and ensure the probes do not land or crash by accident.

Another contactee referenced by Timothy Good was Sidney Padrick. Padrick affirmed Sir Peter's story that the visitors were here to observe. He was told (amongst other things) that the visitors' homeland had no sickness, no crime, no police force, no money and had a strong service-to-others ethic. Sounds like an idyllic world far removed from Earth's experimental training-ground-for-souls environment.

In Gordon Cooper's wide ranging book *Leap of Faith* he talks extensively about UFO's, extraterrestrial existence and such matters as "interdimensional communication". This form of communication was revealed to him in the 1970's during meetings with a lady called Valerie Ransone, who claimed to have a direct line of communication with extraterrestrial sources who were assisting in the passing of technical information to help beneficial development on Earth, such as development of clean energy. Cooper discussed Tesla technology in his book, such as Nikola Tesla's undeveloped patent in 1900 for wireless energy transmission, which Ransone hoped to develop through her ability to receive signals from advanced civilisations not of this earth. Perhaps Nikola Tesla, the inventor of the radio wave transmission and the alternating current electrical supply system, amongst many other inventions, had received similar communications, as his inventions and discoveries were so far ahead of his time at the start of the twentieth century.

Cooper became convinced of the truth of Ransone's ability to communicate with unknown sources through telepathic means. This conviction was, in particular, due to a communication from her sources that she passed to him concerning a fault, which no one had noticed, in the cooling system of the space shuttle that was about to be launched. The fault was duly found and corrected. If the fault had been left uncorrected the consequences would have had devastating. However, the far-ranging plans for new technology using Tesla's principles were not to be realised as financing could not be obtained. Ingrained institutional interests would not allow such developments, even if seemingly beneficial to the human race and the environment.

The revelations of Adamski: George Adamski, who wrote the mass-circulation paperbacks *Flying Saucers Have Landed* (1953) and *Inside The Space Ships* (1955), brought the phenomenon of extraterrestrial life to the attention of the wider public. He was told that alien bodies die just like our earthly human bodies, but the spirit does not die. This goes on evolving. Adamski was the most high-profile contactee from 1946 for a decade or so after he revealed extensive contact with spaceships from the planet Venus and meetings with a Venusian by the name of Orthon. He photographed the space ships, and also claimed he flew in one.

The revelations of Billy Meier: For all of Adamski's incredible revelations, it was the historic and astonishing case of Swiss citizen Billy Meier which took the reality of extraterrestrial contact to new levels, at least so far as the public was concerned, and his case still remains the most intriguing. His evidence of contact from 1975 includes close-up photos and 8-mm film footage of extraterrestrial craft taken over many years which no one has successfully 'debunked'. Like Adamski, his contacts looked similar to humans from Northern Europe. He called his contacts "Plejaren" and stated that the Plejaren homeworld was called Erra. It is stated as being located inside the DAL Universe, a sister universe of our DERN Universe, about 80 light years beyond the Pleiades, an open star cluster. Many have written about the Meier case, including a reasonably favourable review by Gary Kinder in his 1987 book *Light Years*, and a less favourable review by Kal K Korff in his 1995 book *Spaceships of the Pleiades*. For many, the Meier case is proof that "Contact is Real".

Many of the 'contactees' were told how the spaceships propelled themselves through space, some explanations of which were very complex, but the main theme is the nullification of gravity. This allowed infinite speeds. Advanced systems were in place to maneuver the craft so no accident could generally occur, though the crash that occurred at Roswell, described previously, allegedly occurred because the powerful radar in New Mexico interfered with the controlling mechanism of the crafts. As revealed earlier however, it may have been an "intentional" crash.

The revelations of Courtney Brown: Finally, Courtney Brown states from his research through remote viewing that there *is* an entity called the Galactic Federation which governs much of the galaxies. However, this is what he terms a "subspace" organisation, in other words non-physical, as it would be essentially impossible for physical beings to govern a galaxy. The reason is that physical beings are temporary creatures who participate in the physical world only for brief periods of time. Persons who govern a galaxy are obviously required to have memories that are much greater than a lifespan and are everlasting, so that decisions can be made with the benefit of endless millennia of experience. Thus, the Galactic Federation are non-physical beings. As Brown says, "It could never have been otherwise".

The 'space brothers': The point of the work undertaken by the so-called "space brothers" who contacted selected individuals, particularly during the fifties, was to circumvent the increasing government and military secrecy which was restricting all information on extra-terrestrials reaching the public. Notwithstanding the ridicule encouraged by official government attitudes to alien existence, the 'contactees' did make their experiences known to the wider public, writing books and appearing on radio and television. Their message slowly permeated the mass consciousness as reports of these contacts became widely known.

Change is coming

Carl Johan Calleman says in his important book *The Mayan Calendar and the Transformation of Consciousness*, that the ending of the Mayan Calendar in December 2012 ended a long period of developing consciousness of Earth's population. He considers that positive changes are coming, attitudes are being transformed by the changing energies, and old ways of thinking and relationships that are outdated ending as a result of the changing energies of Earth. He says that for those who are not able to embrace the changing consciousness that is arriving, and attempt to fight or block change, conflict is inevitable. Ultimately, change cannot be stopped.

Something in society is certainly changing, but is it positive? Whilst the majority of humanity want to see positive change, it may be that certain global government, authorities, or possibly elite personnel above government level, know what that 'something' is, and it is far from a positive change.

In *The Uninvited*, Nick Pope's review of the alien abduction phenomenon as of the year 1997 (though little has changed since that date), he raises the possibility that contactees may have been the victims of a terrestrial hoaxing and disinformation campaign. It is clear that military and intelligence agencies are taking an increasing interest in the alien and UFO phenomenon, and would have a motive for discrediting ufology in general. Associating alien contact with stories that can attract ridicule is clearly helpful in diverting public interest from a serious subject. Contactees may also have simply been a product of the post-war concern of many against nuclear proliferation and pollution. The message did find a vast audience whatever the motives.

Was the negative change in the public's perception of 'alien contact' from the 1970's onwards a consequence of the change in society? As society had become increasingly impersonal, anti-social, self-serving, violent and polluted, did the reporting of alien contact also become more negative? Whatever the reasons, as will be revealed from the reviews in the next chapter, the change that is coming on Earth could be radical, and perhaps those coming changes will see the end of the human race as we currently know it.

Is the change that may be coming a 'final act' of extinction by aliens in response to humanity's aggressive tendencies and the concern at what humanity may do on a wider galactic stage with its rapidly developing new technology?

Where It All Ends: Abduction - The Alien Agenda and Hybridisation

This chapter is where the darkness of extraterrestrial existence really becomes apparent. Further, if the books reviewed in this chapter have any veracity, potentially we could even be looking at the end of the world as we know it.

Alien abduction has now been heavily investigated and extensively reported and written about, but the starting point for those with limited knowledge of the subject must be *The Threat* by Dr David Jacobs, subtitled *Revealing the Alien Agenda*, written in 1998 and considered a key work on the subject.

Although Jacobs (recently retired) worked as a professor of history at Temple University, Philadelphia, his true focus was as a UFO researcher, and his dissertation for his doctorate in intellectual history from the University of Wisconsin in 1973 was on the UFO controversy. By the early 1970's, the UFO research community had built up a huge data base on UFO movements, including such matters as their effect on the environment, automobiles, electrical equipment, animals and humans. But the UFO's were not making contact with the general public and their motive for entering Earth's airspace remained a mystery.

However some humans were reporting abduction experiences, and over time some common features of these abductions were emerging: paralysis, physical examinations, telepathy, amnesia and little grey beings with large black eyes. These alien creatures seemed to be able to 'switch off' nearby people (in other words, rendering them unconscious or immobile) while abducting others from the same household. Solid proof of these events was, however, hard to pin down. Jacobs decided to investigate the phenomena further because from his viewpoint abduction might be the key to unlocking the UFO mystery as 'abductees' actually went *inside* the UFO crafts.

What he ended up discovering from over 700 regressive hypnotherapy sessions with 'abductees' at the time of writing his book was that the aliens were not visiting Earth for the benefit of humans - far from it. The aliens were on Earth for very sinister reasons. They had a reproduction agenda, the main focus of which was what Jacobs terms 'The Breeding Program', in which the aliens collected human sperm and eggs, incubated foetuses in human hosts to produce alien-human hybrids, and caused humans to mentally and physically interact with these hybrids for the purposes of the development of the hybrids. Aliens also appeared to use implants to monitor human thoughts and movements, particularly those humans who were being used to incubate foetuses or were being used as part of the reproduction agenda.

Abductees have often reported the use by the aliens of a 'mindscan' where the aliens stare into abductees' eyes and stimulate the brain. Once the alien connects into the neural pathways, he has absolute power and control over the mind and body of the abductee. Indeed some abductees have reported that memories and other information related to the abductee are recorded and then transferred into the minds of hybrids so they can learn how humans live and feel.

The Breeding Program would not be possible without the alien control over the human mind, and the abductees have stated that they understand that their private thoughts are not their own and that they can be 'tapped into' and manipulated.

Secrecy

Above all though, secrecy is vital in relation to what the aliens are doing, as without it the alien agenda could not succeed.

This agenda of secrecy appears to have been very successful, as Jacobs reports that most abductees, even those who have had a lifetime of abduction experiences, remain unaware of what has happened to them. Abduction cases studied may be merely the 'tip of an iceberg', with most experiences never recalled, forgotten forever. Jacobs suggests that this may be because the aliens store the abduction events directly in the abductee's long-term memory system, bypassing short-term memory and preventing the triggering mechanism that allows for its reconstitution. Hypnosis restores the trigger that allows the memories to come forth.

Aliens do not erase the memories entirely as there appears to be a future purpose for the memories, which will be revealed below. A further aspect of the secrecy, as mentioned earlier, is the 'switching off' of others nearby to the abductee(s), and then the apparent 'cloaking' of the aliens and the abductee(s) during the abduction process as the actual 'beaming up' to the craft is never seen.

Why so much secrecy? As Jacobs makes clear, it is to ensure the success of the Breeding Program. The early growth of a foetus inside the womb of a human woman must be protected.

Even with all the secrecy however, we know something has been going on: There are marks on the ground where the craft came into contact; the crafts often leave radiation traces; abductees experience actual verifiable 'missing time'; and some abductees have physical scars and conscious memories of the experience.

Genetic harvesting and hybridisation

Jacobs goes into immense detail about the hybridisation of the human to produce *homo alienus*. The hybridisation process goes through various stages, under which the hybrid becomes increasingly human by about the fifth offspring from the initial reproductive process. The late-stage hybrid could essentially pass for a human, but has alien capabilities such as their extraordinary mental abilities. In particular, these late-stage hybrids can have intercourse with humans. It seems that the time when hybrids can live among humans is close.

So what does this all mean for humans? Jacobs reports that the aliens continually refer to the future, a time when change will occur. This will be when the hybrids are able to fully integrate into human society. Then, "people will be different". Jacobs leaves us with a frightening reality, a controlling hierarchy within society where the order of control will see insect-like aliens at the top of the tree, followed by other aliens such a reptilians, tall greys, and small greys, then hybrids, abductees and at the bottom, non-abductees.

According to Jacobs (and others) civilisation is about to undergo a rapid change not of our human design. Jacobs ends by saying that when he was a child he had a future to look forward to. Now he says he fears for the future of his own children.

The viewpoint of Dr Jacobs is reiterated by Dr Karla Turner in her classic 1994 study of the conscious recollections of eight abductees entitled *Taken*, which was republished with a new introduction in 2013. A prominent theme from Dr Turner's book is genetic harvesting from humans to create a new (hybrid) human species.

David Jacobs and Karla Turner are both aware that it is through the abduction phenomenon that we come most urgently face to face with the alien presence: Abductees actually interact with a variety of different aliens and enter into their world, which is the UFO craft they operate from, their 'home' when travelling. As Turner says, "All of the UFO photos in the world tell us nothing compared to the words of those who have encountered the alien force in their lives and those of their families". It is clear we still do not really know for sure who these beings are, and why they are visiting our planet.

The evidence suggests, according to the books reviewed, that there is no spiritual or enlightened agenda of these aliens. If these aliens were "angels", at least angels of the sort humans understand to be interacting with us for our benefit, then why would they be performing unwanted rectal probes; arranging sexual contact against the individual's wish; undertaking intrusive physical examinations which leave bodily damage, punctures, bruises, implants; teleporting the bodies; controlling minds and creating illusion in the minds of abductees; programming abductee minds to suit the aims of the aliens; suppressing memory to limit the possibility of the abductee becoming aware of what actually occurred in an encounter with the alien; and much more?

Due to deception, secrecy and illusion employed by the aliens, abductees only feel comfortable revealing their encounters when there is actual physical evidence such as body punctures, scoops, patterns of bruises, marks on the ground and so on. Yet, studies by mental health professionals show that the people who make these reports are clearly sane and the information is derived from personally experienced trauma.

Turner's interviews with her abductee subjects constantly return to the idea of coming destruction and change on Earth. She states that the aliens are currently (as they did in the past) carrying out genetic alterations of the human species which will ultimately supplant us, and is being done for the benefit of aliens and not the human race. Her view is that they want us, as a race, to be so afraid of this upcoming destruction that when they show themselves openly and offer to save us in some way, we will be willing to take their help, even if it means giving up our sovereignty and surviving under subjugation. Turner was told telepathically that we should never allow this to happen.

The minds of humans are very open to suggestibility, as will be addressed further in the section on mind control in the chapter on Global Control of Humanity. In the chapter on global control, David Icke makes clear in his book on the "global conspiracy" that methods are used to persuade us to willingly give up our human rights, free will and sovereignty. Whether employed by elite humans or aliens, conceding our sovereignty and free will can never be acceptable.

Perhaps Earth is an experimental planet, a cosmic zoo where humans and animals are used for genetic experimentation and species alteration. Are humans now waking up and starting to see through the apocalyptic visions and illusion of the aliens? As Turner suggests, perhaps awareness by humans of their abduction experiences helps to evolve their consciousness and become more empowered. As one of Dr Turner's interviewees ("Angie") said, she has prayed for help from the good sources in the universe, but no help has come. She therefore decided to take the matter into her own hands and use her own God-given reasoning faculties, reasoning and prayer in deciding how to respond to the aliens, and she ended up saying this: "I am a human being, and we humans do have feeling and rights!"

Time travelling

In his 1998 book entitled *Time Travellers from our Future*, Dr Bruce Goldberg took a slightly different view of the alien abduction agenda. He stated that, from years of hypnotic regression/progression of patients, he had concluded that the abductions of humans by aliens were happening for two reasons: firstly, to solve massive fertility problems that humans will have in the future; and secondly, to ensure the survival of the human race by assisting in humanity's spiritual growth. Under hypnosis, the patients provided information that the aliens who carried out the abductions and reproductive examinations and experiments have come back to Earth from a future time (from between 1,000 to 3,000 years ahead).

Dr Goldberg takes the view that the alien time travellers use some form of hyperspace engineering and enter a wormhole (a connection between white holes and black holes in space) to transport them back in time. Goldberg says that there are at least five 'parallel universes' with their own timelines which can be used to move a person from a negative timeline to a more positive one. For instance, on one timeline there may be a nuclear war, which is undesirable for the people on that timeline, so a person can shift to a new timeline and avoid the war by the simple process of changing their conduct.

In *Glimpses of Other Realities*, Linda Moulton Howe reported on an extraordinary hypnosis session in 1994 with James Peniston, a US airman who witnessed the "British Roswell" incident at RAF Bentwaters fourteen years prior to the hypnosis session when a UFO craft landed outside the base. Peniston had direct contact with the aliens. Amongst other things, Peniston stated under hypnosis that the craft had a problem and needed a place to stay while the craft repaired itself. He stated that the aliens were time travellers from the future and that they had been coming to Earth for at least thirty to forty thousand years, to try and resolve a reproduction problem, to sustain their children in the future. Linda also received a report relating to the Cambodian incident referred to in an earlier section that non-humans were using the cover of wars such as Vietnam to harvest tissue and genetic material from animals and humans for their genetic experiments.

Peniston stated that the future children of Earth will be hairless, humanoid bodies with pale skin and large eyes to take in more light as the future earth will be different. These time travellers can generate enough speed in their craft to go back forty to fifty thousand years but not much further back in time than that.

Linda explained in her book the mathematical theory that if matter approaches the speed of light, time appears to slow down to an external observer. The closer you got to the speed of light, the slower time would go. If you reached the speed of light then time would stop. Then if you take the theory further and exceed the speed of light, time would reverse. This would appear to be how the alien 'time travellers' moved backwards in time. An example of time travel that people may be able to relate to is the 2006 film *Déjà Vu* where the concepts of time travel were explained in a real life environment.

This view of time travellers from the future has been addressed in significant real time detail by Dr Dan Burisch, a micro-biologist working at a secret location ("S-4" in the Nevada Desert) with an alien being called "J-Rod". Dr Burischwas interviewed by Paola Harris in her book *UFOs: How Does One Speak to a Ball of Light?*. Dr Burisch stated that "J-Rod" was the short form name given to the creature who had come from a base in the Aquarius Constellation: Gliese 876C, some 52,000 years ahead of the present time. His craft had crashed at Kingman, and he was one of three creatures, the other two were from about 45,000 years ahead in the future. Communication was by electromagnetic entrainment, which is the method used during alien abductions. It would allow the creature to bring Dr Burisch into his world, interact and participate in what he was thinking and feeling, including life on his home planet. J-Rod had returned from the future to change a catastrophe which had happened in the past (allegedly in 2012 when there had been a split between spiritual and technological beings) and move to a different timeline. He was returned to his future time utilising a 'stargate' device known as the "Looking Glass", which was depicted in the 1997 film *Contact*. This device was constructed in the Egyptian desert, from original diagrams for stargate devices based from ancient cylinder seals. Dr Burisch gave significant detail about the physical workings of the Looking Glass in his fascinating interview. Al Bielek stated in his discussion with Ms Harris that the Looking Glass device was being used by certain governments almost as a toy to travel through time and change history. He stated that the time field is a closed loop, so that if you go far enough forward in time, you will wind up crossing over and going into the past, at the infinity point.

The writings of an actual hybrid

The apocalyptic view of Dr David Jacobs, and infor-mation given by aliens to many of the abductees, as can be seen from Karla Turner's *Taken*, is echoed by James Walden in his 1998 book entitled *The Ultimate Alien Agenda*, although like Dr Goldberg, James Walden did have a more optimistic view of coming changes on Earth. The book is a personal account of his interaction with aliens who abducted him. He was able to recall the experiences through working with an experienced hypnotherapist, Barbara Bartholic. He consciously witnessed an alien in his bedroom, and through his work with Barbara Bartholic ended up recounting an incredible story.

What he discovered was that *he was a hybrid*. He had started out in a glass tube, and then his embryo was brought into Earth's dimension and implanted into his mother's womb. His mother never knew about the alien implantation that resulted in his birth. His foetus was part human and part alien. Prior to Walden's mid-forties however, he had no recall of alien abduction, even though it had gone on all his life.

During the abduction experiences he recounted under hypnosis, he stated that the aliens could move his body through walls, ceilings and so on. This was done by transforming the molecular structure of his body by lowering the temperature of the physical body and transporting his etheric ("permanent" or "natural") body away for experimentation, where he was used for reproductive purposes, re-programming and upgrading. Walden stated that his other (alien) body would return to another dimension when his human life ended. The masters of the hybridisation program were reptilian in origin, and these beings had colonised Earth many years ago. This is why ancient art depicted regal reptilian figures. Walden noted, from further research he undertook following his work with Barbara Bartholic, that the art, myths, and stories of most ancient civilisations maintain that dragons, serpents, and lizards have long co-existed with humans, and that many statutes, drawings and carvings portrayed the reptiles as winged beings or gods who had descended to Earth and become involved in the lives of humans (see the earlier review of RA Boulay's book *Flying Dragons and Serpents*). The Gnostic writings also intrigued Walden as they portrayed the role of the serpents on Earth and were not included in the bible as they conflicted with the story the church authorities of that era wished to portray.

Conclusion – replacement of the current human

It seems that the human hybrids are monitored by the aliens for the coming event on Earth when hybrids will replace the current human. That event is coming soon, it would appear, as the hybridisation process is almost complete. Walden was of the view that the ultimate goal of reptilian intelligence was to populate the Earth with integrated human hybrids – in other words, people with human minds and bodies who could simultaneously manifest multidimensional reptilian intelligence. Human hybrids are the carriers of the reptilian intelligence and within a generation or so the pure human species will dissolve and be replaced by the human hybrid. Walden went on to state that "pure humans will not exist in the future".

This brings to mind a scene from *1984*, George Orwell's dystopian masterpiece, where the hero of the story, Winston Smith, is tortured and informed about his likely future:

"Never again will you be capable of ordinary human feeling. Everything will be dead inside you. Never again will you be capable of love, or friendship, or joy of living, or laughter, or curiosity, or courage, or integrity. You will be hollow. We shall squeeze you empty and then we shall fill you with ourselves".

Is this the fate of today's human?

Alien Viruses

If the current human is to be phased out, how will this be done? A scenario of possible annihilation of Earth's human population is considered by Dr Robert M. Wood, in his book entitled *Alien Viruses: Crashed UFOs, MJ-12, and Biowarfare*, which he wrote in association with Nick Redfern. Dr Wood sets out a logical basis for planetary takeover by aliens. It is clear from other reviews in this book that there are alien groups who may well be far from benevolent in their attitude towards the human population of Earth, and may be looking at suitable ways to reduce or eliminate the population.

The introduction of viruses that all humans, or perhaps certain categories of humans bearing distinctive genes, are unable to defend themselves against is plausible, as it would avoid excessive property damage, environmental damage and planetary turmoil, and could be carried out secretly without harmful warfare.

Consequently, as Dr Wood sets out in detail, activities such as the on-going cattle mutilation program might have sinister motives in relation to the understanding by the aliens of earth-based genetics. This mutilation program may be for the purpose of assisting in the development of lethal pathogens for removing *homo sapiens* from the planet.

Viruses may well currently be under experimentation, such as CJD, Aids, Ebola, various flu viruses and so on. Once it has been established over a number of years as to how different viruses affect the population, the aliens might be well placed to take decisive action. It is possible that certain humans could be saved from a near-total elimination virus by a vaccination that could be developed as part of the study and by development of suitable viruses.

As Dr Wood states, anyone wanting to understand how to vaccinate against new diseases would find that studies of infected cattle might be helpful. He points out there is significant evidence that alien craft (as well as covert military craft) are mutilating animals, though this is mainly cattle.

The process of developing lethal viruses has most likely been going on for a long time, since at least the first UFO crashes after the Second World War. Dr Woods and Nick Redfern go into immense detail concerning recent alien history, including dead alien bodies held in cryogenic storage within secure official facilities, the vast alien cover-up story since the Roswell crash in 1947, and the uncovering of classified documents and whistleblower testimony.

This leads to the conclusion that a release of extraterrestrial-originated viruses may not be so far away (if it hasn't already occurred). And, most certainly, that we are not alone in the Universe, and that our unearthly visitors most probably pose a direct and very lethal biological hazard to the human race.

It is revealed that in recovering the alien bodies from the Roswell crash in 1947, certain technicians experienced alien virus contamination resulting in death. In the

Majestic Twelve 1st Annual Report of 1952 revealed in the book, it is indicated that samples extracted from the alien bodies found in New Mexico had yielded new strains of retro-virus not understood but extremely lethal and which gave promise of the ultimate biowarfare weapon.

The best publicised case of human death from an alien virus was the "Varginha" case in Brazil which ended with the rapid death of a Brazilian man following contact with an alien species. Paola Harris interviewed AJ Gevaerd, editor of Brazil's "UFO Magazine" concerning the death of a policeman following the discovery of an alien creature who had been on board a crashed UFO in January 1996 in Varginha, Brazil.

Gevaerd stated that Marco Eli Cherese, a local policeman, participated in the capture of an alien creature on 20 January 1996, and during the capture had touched the creature directly with his left arm without gloves. This had contaminated him in some way leading to his rapid death. The doctor in charge of the case, Dr Furtado, told Gevaerd that some unknown virus penetrated inside his organism and deprived him of his immunity system very rapidly. This is exactly the sort of virus that could cause planetary extinction if unleashed on a wider scale.

It can be readily concluded from both the evidence in Paola Harris' book and Dr Woods' book that alien viruses pose a real threat to humankind.

Will this be the silent and swift method for ending humanity?

PART 5

Political Impact,
Control and Destruction

Exopolitics: The policies and attitudes of governmental authorities and the military concerning the existence of extraterrestrial life

At the present time, the existence of an extraterrestrial presence is undisclosed. The leading work on the important subject of "Exopolitics" (the extraterrestrial presence, what it means to humans, and how we can manage this reality), is Michael Salla's 2009 book entitled *Exposing US Government Policies on Extraterrestrial Life*. It is an extraordinary source for verification of extraterrestrial presence, as will be shown.

Exopolitics follows on naturally from the Serpo project information, and is a new discipline dedicated to studying the political implications of extraterrestrial life and presenting the information to the general public with a view to producing 'democratic' decision-making concerning extraterrestrial life.

Government agencies and secretly appointed committees have been responsible for most policies and decisions taken to deal with extraterrestrial life, and the wider public have been excluded from this process. As Michael Salla points out, allegedly 'the best minds' have discussed problems supposedly too complex and disturbing for the general public.

The main justification for non-disclosure of evidence confirming the existence of extraterrestrial life is that US policy makers believe that the general public is simply not ready. The main purpose behind the discipline of exopolitics is to prepare the general public for the truth concerning an extraterrestrial presence. To some extent, public acclimation is already being undertaken through blockbuster movies based on 'science fact', such as *Close Encounters of the Third Kind*.

Salla's book is a veritable "who's who" of experts and personalities associated with the alien phenomenon, and for those who are new to the subject, this is a good place to start.

Evidence

Because the existence of extraterrestrial presence on Earth is undisclosed and hidden in secrecy, Salla provides evidence much like one might do in a court hearing to overcome the lack of a "smoking gun". He therefore discusses witness testimonies in relation to five key areas: 1. A meeting between President Eisenhower and extraterrestrial visitors; 2. Official agreements with extraterrestrial visitors; 3. The sighting of extraterrestrials at classified facilities or operations; 4. Official communications with extraterrestrials; and 5. Extraterrestrials living and/or working among us with official approval.

Substantial evidence is produced, but whistleblower and other similar evidence is often discredited in one way or another. The evidence can therefore either be viewed as 'science fiction' or 'science fact', or simply fascinating story telling depending on one's own leanings. However, one of the more interesting stories (within items 3-5 above) revealed in Salla's extensive writings is as follows:

In 1987, an alleged whistleblower by the name of Thomas Castello released photos, papers and a video which appeared to be evidence of a joint US government/extraterrestrial secret underground base at Dulce, New Mexico. These papers became known as the "Dulce Papers" and described (amongst other things) genetic experimentation, development of human-extraterrestrial hybrids, use of mind control through advanced computers, and cold storage of humans in liquid filled vats. The Dulce Papers were circulated on the internet and were incorporated into a book called the Dulce Wars. It backed up whistleblower and investigative work, but a few years after revealing the information the said Thomas Castello disappeared.

Castello claimed that there existed a seven-level underground facility at Dulce which housed both humans and extraterrestrials. There were allegedly four extraterrestrial races residing at Dulce, the standard 'short' 'Grays' from Zeta Reticulum; tall Grays from Rigel, Orion; a reptilian species native to Earth; and reptilians from the Alpha Draconis star system. The Earth based reptilians were allegedly led by a winged reptilian species described as the Draco, and the short Grays were subservient to the Draco. The different projects at Dulce involved reverse-engineering of extraterrestrial technology, development of mind control methods, and genetic experiments involving cloning and creating human-extraterrestrial hybrids. Reports included abuses of abducted humans by extraterrestrials which may have led to some sort of an underground battle between human military and aliens, in which humans used weapons and aliens used directed mind energy.

Abuse of humans

This whistleblower was not the only source that revealed information about alien abuse of humans. David Icke also mentioned in his book *Children of the Matrix* that there are various alien groups who live underground in known locations, often in America, all of whom are not benevolent, and who have committed and continue to commit atrocities to obtain human genes and continue the on-going hybridisation programs.

Colonel Philip Corso had also referred to extraterrestrials who he said were abducting civilians, violating airspace and destroying aircraft sent to intercept them and were therefore a direct threat to US national security.

Researchers such as Dr David Jacobs believe that the 'Grays' have a covert plan to takeover human society by engineering a superior hybrid race, and have provided case studies of shocking abuse in alien abductions (see the chapter on abduction); yet, Dr John Mack believes that the visitors intend to blend together the best features of extraterrestrials and humanity. The late Dr Carl Sagan often warned that we should refrain from announcing our existence by transmitting radio signals into space "because we do not know the intentions of a superior galactic society". Maybe he knew more than he said.

Underground bases

It has also been claimed by Richard Sauder in his 2001 book *Underwater and Underground Bases*, that the underground infrastructure is so vast that travel underground utilises advanced technology such as a high-speed rail link. Bill Hamilton is quoted by Salla as stating in *Cosmic Top Secret* that "there appears to be a vast network of tube shuttle connections under the US which extends into a global system of tunnels and sub cities." Sauder states that the technology to construct underground and underwater facilities and tunnels has been around for decades. A publication thirty years ago by the US Army Corps of Engineers stated that "since adequate technology is available to construct hardened underground facilities under virtually any ground conditions, the main constraint in construction projects remains economic viability rather than technical feasibility". Bearing in mind the size of US black budgets (as discussed in detail below), financial feasibility concerns are unlikely to exist. We know that highly complex underwater train links such as the England-France tunnel have been constructed, and Sauder talks of the technical feasibility of the use of Maglev trains (being trains that can travel at rapid speeds suspended on a magnetic field as they travel without friction from rails) for underground travel across vast distances. Moreover, Sauder points out that the technology to construct facilities and tunnels *beneath* the ocean floor has also existed for decades. The big question is what are these vast underground and under-ocean facilities being used for? Because of the lack of transparency and secrecy, it is not unreasonable to conclude that they are not benefiting the mass of humanity. It seems that a secret world exists which is inaccessible to the masses.

Black budgets and secrecy

No one can say whether all, some, or none of the information is true, because nothing is proven, everything is denied, and disinformation campaigns are often used in which the factual process is exaggerated by selectively distorting parts of the 'truth', leaving the reader with no idea what to believe, if anything. The cover-up is very thorough. Intelligence officers, military personnel and pilots are all silenced under the official secrets acts around the world. What is certain is that the public will never know the truth about aliens, or of alien abuse, or whether the public is just subjected to negative publicity, until transparency of military and government activities occurs. That appears to be a key function of "exopolitics": to assist the process of bringing such information into the public domain and preparing the general public for the reality of extraterrestrial existence.

But if such extraordinary structures referred to above *did/do* exist, and extraordinary work such as reverse engineering of alien technology *was/is* being undertaken, it would require vast sums of money, beyond conventional budget approvals which are transparent and are able to be accessed by the public. How would such secret projects, which might require trillions of dollars to build and maintain, be funded? It is said that such projects are funded through a mechanism known as "the black budget".

Salla states that the "black budget" funds a covert world of unaccountable intelligence activities, covert military/intelligence operations and classified weapons programs. The black budget allows intelligence activities, covert operations and classified weapons research to be conducted without Congressional oversight. The justification used is that oversight would compromise the secrecy essential for the success of such 'black' programs. These black programs are typically classified as "Special Access" or "Controlled Access" programs and have a security classification more rigorous than the 'top secret' classifications for most governmental agencies. Such programs are known only by those with a 'need to know'.

The last thing that people who oversee secret projects would want is for the mass of humanity to get to hear about (or worse still, access to) the development of expensive technology which will take humans through space to the world of living extraterrestrials. The masses might help with the funding, but these projects are most definitely secret. Benefits from alien-derived technology are not for sharing.

Fortunately for those utilising black budget funding, the 1949 CIA Act makes such black budgets entirely legal. The relevant provision states that "any government agency is authorised to transfer to or receive from the [CIA] such sums *without regard to any provisions of law* limiting or prohibiting transfers between appropriations. Sums transferred to the [CIA] in accordance with this paragraph may be expended for the purposes and under the authority of sections 403a to 403s of this title without regard to limitations of appropriations from which transferred." According to L. Fletcher Prouty, who was Chief of Special Operations for the Joint Chiefs of Staff from 1955 to 1964 where he was responsible for military support of the CIA's covert worldwide operations, this led to the creation of a "power elite" that essentially formed an unelected *secret government* that used visible government institutions and personnel for its goals. Senator Inouye also stated at the 1987 Iran-Contra Senate hearings that "there exists a shadowy government with its own air force, its own navy, its own fund raising mechanism, and the ability to pursue its own ideas of the national interest, free from all checks and balances, and free from the law itself".

As stated above, the idea was ultimately to create a global secret government. This secret government did not acknowledge the existence of national sovereignty. In order to control the world effectively, it had to control global corporations that were transnational, who themselves controlled the things that make society function: food, water, air, transport, and so on. This process was started during Prouty's time and is effectively complete today, as William Cooper indicated in an earlier chapter. For more information about this subject, see the later chapter entitled "Global Control of Humanity".

The CIA Act, in fact, completely bypassed the transparency of elected officials as it stated that "the sums available to the [CIA] may be expended without regard to the provisions of law and regulations relating to the expenditure of Government funds; and for objects of confidential, extraordinary, or emergency nature, such expenditures to be accounted for solely on the certificate of the [CIA] Director." Thus, the CIA Director can simply utilise, legally, taxpayer funds for secret projects if they fall into a category which is confidential, extraordinary or emergency. This would seem to adequately cover extraterrestrial invasion. The public would be suitably terrified to hear of any such potential threat, so the enormous expenditure would have to remain highly confidential. The vast numbers of projects related to the extraterrestrial threat include reverse engineering of extraterrestrial vehicles, developing advanced technologies based on information supplied by extraterrestrials, and the creation of deep underground bases, amongst many other things.

Until greater disclosure occurs, no view on the rights or wrongs of the 'black budget' can be easily offered. We need to know more about the types of aliens faced and the threat or benefits. If there is any truth in some of the whistleblower information, humans need a lot of protection, which is why it appears that 'black projects' exist for an advanced planetary defence system capable of ensuring the sovereignty of major countries and humanity in general. The idea that extraterrestrials were here to offer 'spiritual development' was considered too risky a choice rather than defence capability. The problem is, can humans trust the people who are defending their sovereignty when their activities are so shrouded in secrecy? This is where the discipline of "exopolitics' comes in.

There have been a number of political options as to who the authorities should deal with: (i) extraterrestrial competitors for Earth's resources (including human resources), (ii) extraterrestrials who were spiritually and ethically advanced and offered non-military forms of assistance for advancement of society, or (iii) a group who merely monitored Earth's activities and were therefore less bothersome.

A number of writers have said that decisions were made (deriving from the above options) to create a strategic alliance with the Grays and the Reptilian extraterrestrials who had advanced technology to trade but in return required access to humans for genetic and biological enhancement purposes.

The use of HAARP

Those extraterrestrials with whom deals had not been done could be excluded from entering Earth's atmosphere through the use of the High Frequency Active Auroral Research Program (HAARP) which can be used as a particle beam projector capable of generating a global shield. High speed ions are projected into the magnetosphere, thereby destroying the electronic systems of any craft entering Earth's atmosphere. The global shield created by HAARP could cause alien crafts to crash, and these crashed crafts could then be acquired for technology transfer by the relevant authorities or intended beneficiaries. An interesting thought is that perhaps the military and the current human controllers of Earth have now reached somewhere near parity with the extraterrestrial technology, after playing 'catch up' for half a century.

In brief, the HAARP system is designed to manipulate the ionosphere, a layer which begins about thirty miles above Earth. The HAARP device on the ground is a large field of antennas designed to work together in focusing radio-frequency energy for manipulating the ionosphere. The radio-frequency energy can be pulsed, shaped, and altered in ways never possible before. The key point is that the device can power from Earth onto the ionosphere one billion watts of energy, which enables the device to be used (amongst other things) for "earth penetrating tomography" (looking through layers of the Earth to locate underground facilities or minerals), communications with deep submarines, to manipulate the communication of other people, over-the-horizon radar, the alteration of weather, to create artificial plasma layers or patches in the ionosphere, change the mood and mental state of people, particle beam weapons, and as an anti-satellite weapon. The effect of HAARP will be examined in greater detail in the chapter entitled "Global Control of Humanity", and in particular a review of information on HAARP from the 1995 book by Nick Begich and Jeanne Manning entitled *Angel's don't play this HAARP: advances in Tesla technology*. HAARP may well be assisting in the rapid destruction of the planet's protective layers and the attempts to correct this through atmospheric spraying.

Conclusion

In the final analysis of extraterrestrials and political "spin", at the present time the trend is towards the idea that extraterrestrial activity is hostile, which is the justification for allowing trillions of US Dollars to be spent on black budget activities. This has been due to an increase in reports of abusive behaviour and away from the previous friendly 'space brothers' of the 1950's to 1970's. These abusive abduction reports may be either extraterrestrial or perhaps human military-intelligence abductions ('milabs'). In the case of milabs, it is said that these are military counter-intelligence operations, which are undertaken in order to attempt to learn the alien agenda. Through a combination of chemicals, hypnosis, and possibly other technologies, attempts have allegedly been made to extract the memories of the victims of alien abduction, and then wipe them to avoid detection of the milab operation. In his book *The Threat*, Professor David Jacobs argued that alien abductions were occurring which involved sexual and reproductive operations designed to produce a hybrid species which will replace the current version of homo sapiens. This abduction and genetic manipulation activity has been recorded by Jacobs as being cruel, painful, sadistic and emotionally dysfunctional, and is a clear threat to the human race. Whether the abductions were through alien involvement, milabs, or both, the 'spin' is negative for the most part.

A researcher and abductee, Dr Karla Turner, in her book entitled *Taken*, also backs up the information revealed by Dr David Jacobs and James Walden that the alien agenda is not benevolent and that humans are a resource for aliens. Her research indicates that apart from the hybridization agenda indicated by Dr Jacobs, aliens 'harvest' (a term used by Colonel Corso) from humans in a number of ways: emotionally, energetically and, most particularly, physically.

If the abduction research has any validity, it must be asked whether 'spiritually enlightened' aliens would hold humans against their free will. Free will is an essential feature of spiritual development, and this does not appear to be a feature of alien abduction and the alien agenda. Indeed, Linda Porter stated, in an interview with Linda Moulton Howe in *Glimpses of Other Realities*, that "unless a person has been through the trauma of an abduction, they cannot comprehend how these entities can strip the mind and soul as easily as they can strip the clothes off a body. In a blink of an eye, these beings devastate all personal freedom, all sense of privacy, and all prospects for living a normal existence. A person's life is never the same again. He or she is changed forever."

Is there perhaps a secret hybridization program for a post-apocalyptic world? After all, it is clear that there is an on-going storage program for seeds at the Svarlbard Seed Bank for a post-apocalyptic world. Richard Dolan, the well-known UFO researcher, has said that "It is fair to conclude that hybridization is part of the [alien] picture. People are being abducted against their will. [Alien] forces seem to be conducting genetic experimentation and acting on us to create a hybrid species".

The remote viewer Dr Courtney Brown, in his fascinating book *Cosmic Voyager*, has said that the activities of the extraterrestrials, in particular the Greys, do have a long-term benevolent purpose, but that there are some who are rogue and form a troublesome minority in the galactic community. *Cosmic Voyager* is reviewed in greater detail in the later chapter entitled "Planetary Destruction", due to his remote viewing work indicating a likely environmental disaster in the near-term future.

Michael Salla referenced the testimony of Dr Carol Rosin for the Disclosure Project. She stated that she had been present in confidential corporate meetings where disclosure of the extraterrestrial presence would be revealed as a justification for military expenditure, but only after international terrorism failed/ceased to be a credible justification for the vast military expenditures of the US military. These expenditures are vital to major

US corporations that actively service the vast network of ET-related projects.

There are 'threats' which can be utilized to ensure that the expenditures continue, including terrorists, Russians, third world threats, asteroids and finally, the alien threat. It is just a case of how the political 'spin' is directed. The important thing is to shape the public's perception of what extraterrestrial contact will mean.

The media has successfully developed an agenda for dissemination to the wider public which is purposely negative, as can be seen from films such as the block-busters "*War of the Worlds*", "*Independence Day*" and "*Battle:Los Angeles*". The 'alien threat' may be helpful for promoting military expenditure, and maybe this is justified. For members of the public, there is no way to assess the truth; all one can do is listen to the information provided by authoritative sources. But we would all do well to learn as much as possible before First Contact is announced. This will help us all to assess the validity of statements made by the authorities.

The work of "expolitics" will need to be well managed and involve true diplomats at the time of "disclosure", as so much is at stake. However, whilst disclosure is inevitable at some point, it will take an event of monumental proportions to persuade the authorities to reveal the existence of extraterrestrial life, as they have successfully weathered many clear sightings of UFO's and attempts at revelation over more than half a century.

There has been photographic evidence, investigative journalism, whistleblowers and leaks, public confessions, physical evidence, foreign declassifications and public statements, and heavily-documented mass sightings; yet, the authorities stubbornly cloak the subject in secrecy.

Certainly, disclosure will threaten established interests. In this respect, Richard Dolan, a well-respected UFO researcher, and Bryce Zabel have written a projection of what would happen post-disclosure in their interesting book entitled *A.D. After Disclosure*.

But Michael Salla's final, and valid, last comments in his excellent book are: "Understanding and learning about extraterrestrial life and First Contact is the best way in which private citizens can influence 'when' and 'how' First Contact will occur. This will limit the political spin that will be used to shape public perceptions about extraterrestrial life". Knowledge is always the key.

Global Control of Humanity

"The very word secrecy is repugnant in a free and open society, and we are as a people, inherently and historically, opposed to secret societies, to secret oaths, and to secret proceedings. We decided long ago that the dangers of excessive and unwarranted concealment of pertinent facts far outweigh the dangers which are cited to justify it."

President J.F Kennedy in a speech on April 27, 1961 before the American Newspaper Publishers Association

Notwithstanding these words from the late President Kennedy, transparency in the business of Government has not come about. There are good reasons for this, as transparency rarely serves agendas which might not be accepted by the mass of the population. And when those agendas are about control of the population without their knowledge, secrecy is very desirable.

It is becoming increasingly apparent just from reading the daily news that society is heavily controlled. How is this achieved? The methods of the rulers of society have been, and continue to be, brilliant and successful, as William Bramley discussed in detail in *The Gods of Eden*. The key methods of control are summarised below:

1. Breed conflict. Breeding conflict between people can be a very effective tool for maintaining social

and political control over a group of people. The technique, made famous by Machiavelli, is to (a) create conflicts and issues which will cause people to fight amongst themselves rather than against the perpetrator (b) remain hidden from view as the true instigator of the conflicts (c) lend support to all sides of the conflict and (d) be viewed as the benevolent source which can solve the conflicts even though you are the guilty party.

As William Bramley points out, the requirement to stay out of sight was achieved by the creation of secret societies, the masonic lodges, a far-flung network of politically powerful organizations that could create conflict secretly. The Controller race used the secret societies to breed war between humans. This still goes on today. Wars are a perfect tool for making physical survival on Earth an all-consuming chore, because they absorb large-scale resources and offer little to enhance life in return. The idea is to destroy what has been created so that humans are required to start from nothing again. Fighting must be kept artificially alive by creating unresolvable 'issues' that can only be settled by the complete annihilation of one side – and then ensuring this doesn't happen by equalising their fighting strengths. So-called friendly Governments do this all the time.

2. Religious control. Doctrines of fear are very much part of today's religions. The concept of the apocalypse, or Day of Judgment, are rampant in practically every religion, the idea that you must follow specific rules or else you would not become one of the chosen ones, whatever that is, and would

end up in some hell-like place for eternity. This was, and still is, a hugely effective tool for creating fear, control, obedience and misery and at the same time creating funding, as people were required to donate funds to ensure a better life. These forms of indoctrination ensured people would tolerate, and even welcome, unremitting physical hardship, conflict and death. Religious texts and leaders even encouraged the faithful to go out and attack those who held different beliefs. This would ensure eternal salvation. Each religion proclaims it is the only true religion, thus setting in motion more senseless conflict between the believers and non-believers. Religious wars have blighted Earth for millennia and still do. Religious and spiritual doctrines which exalt combat are doctrines which degrade the human race. Violence should have no place in society. The hunting down of heretics, the use of torture such as during the time of the Inquisition, and the warnings against 'false prophets' has led to some of the most bloody fighting between Hebrews, Christians and Muslims. The hangover from the burning and torture which was inflicted on people for centuries for disbelief in a particular religion is still visible today in the instinctive fear people have in expressing any kind of non-conforming idea. Controller mind-control of the highest order. For a detailed analysis of why religion is harmful, one need look no further than Christopher Hitchens' 2007 bestselling paperback *God is not Great*, which has the subtext "how religion poisons everything". Chapter Two, entitled *Religion Kills*, says it all.

3. Disease. Diseases such as the Black Death, and recent virus spreading are likely to be man-made. They create great suffering and conflict. If life starts to progress too successfully for the masses, a wide spread plague will successfully bring this to a halt if a war hasn't. Bramley made the stunning assertion from reports filed at the time of the arrival of disease and plagues in past centuries that plague was preceded by foul-smelling mists and unusually bright lights – the lights of UFO's. Governments today can spread disease easily if they so desire; they have stock piles of viruses readily available. Perhaps the Controllers of that time did too.

4. The monetary system. The creation of today's monetary system which lends itself to economic hardship, through debt and inflation, remains a vital tool for the rulers of society. Governments created the modern money system for a simple reason: It enables them to fight and prolong wars, and control society. Banks have the legal right to create money out of thin air, backed by nothing, and then lend that not-in-existence money to someone against security of a real asset. Mostly they don't even bother to print the paper. But either way it creates debt and inflation, and lowers the value of the money we possess. Of course, should the debtor fail to pay, the real asset owned by the debtor is taken in as compensation. This is easy to manipulate, because the rulers control the money supply anyway. The struggle of the average human for survival in a modern economy where debt creation, inflation which erodes the value of people's money, and

taxation constantly take their toll is exactly what society's controllers desire.

5. Rules and Regulations. Happiness can be easily and effectively reduced through rules and regulations, either through religious laws or state laws. Many religions and nations actively forbid alcohol, gambling, dancing and singing, and limit sexual pleasure, amongst other forms of entertainment used by humans to create happiness in their lives. Religions for instance have expounded the idea that "a man at ease was a man lost". Heaven help a man that was enjoying his life.

6. Political doctrines. A classic conflict was created by the arrival of communism versus capitalism. These opposing philosophies produced the half-century long "Cold War" and much suffering. The idea of 'good' political doctrines fighting 'bad' political doctrines produces inevitable conflicts. Communism for instance was an extraordinary concept where everyone was supposedly equal and encouraged its citizens to accept conditions of social repression and dictatorship. This does not lend itself to happiness. The German Nazi experience created a vicious world war, extraordinary ignorance and obedience to a dictator, based on doctrines of 'good' and 'evil', and it all happened within a time period that many can remember. Conflict through the Nazi activities were stirred up in true Machiavellian style through secret forces, apocalyptic urgings, a paper money banking elite that funded the war through debt creation of non-existent money, and mass exterminations.

7. Media mind programming. The mass population have long been served a diet of only what the mass

media desire you to hear. A small number of elite individuals and corporations control the media outlets, and therefore both the content of the program you watch on television ('programming') and the use of subliminal imagery ensure control over our thought processes. This has been most brilliantly portrayed in long-ago movies such as John Carpenter's 1988 'science fiction' masterpiece *They Live* and Sidney Lumet's 1976 classic *Network*. Sometimes movies reveal more than words can.

In the end humans, however far they rise up the hierarchy of society, are still controlled by the "Controllers". This reality is discussed further below. Life is always going to be precarious until we discover our true spiritual being and essence. What is clear is that the above principles for control over human beings have helped to keep people fighting, suffering, and dying for absolutely no good reason whatsoever.

In literary terms, the starting point for the control of humanity, the physical and mental control that is the basis for a dystopian society, is George Orwell's *1984*, written and published just after the end of the Second World War. Orwell portrays his remarkably accurate vision of a future police state, one where there is no freedoms, no happiness, and control by the ruling party is absolute. As Erich Fromm said in 1961 in his *Afterword* to *1984*, Orwell's masterpiece is "the expression of a mood, and a warning. The mood it expresses is that of near despair about the future of man, and the warning is that unless the course of history changes, men all over the world will lose their most human qualities, will become soulless automatons, and will not even be aware of it." This is essentially the loss of individuality, love and critical thought, and a world where there is no hope for the mass of humanity. As we will see from the reviews below in the section entitled *Elite Control*, this may be the point in time which we have now reached.

The ruling elited

Did the Annunaki really head home when the orbit of their home planet came within range of Earth around 200 BC? Maybe, but even if they did, they left behind their hybrid offspring who, in one form or another, appear to still rule the world.

That is the view according to David Icke in his *Guide to the Global Conspiracy*, a voluminous and persuasive book written in 2007 which is now a leading text on this subject. The book explains how the Annunaki retained their special bloodlines and, through a system of emotionless control, have never let go of the reins of power since they first arrived on Earth all those millennia ago.

The idea of a ruling elite which is of extraterrestrial origin is not a new one, and there is a vast sea of literature out there now which covers the topic, not least William Bramley's *The Gods of Eden*. Bramley's work is an excellent source if you are looking for a well-researched analysis of the origins of warfare, and how this relates to the extraterrestrial being. If you are looking for a "fiction" account of extraterrestrial history and global control, Robert Doherty's extremely readable *Area 51* series should more than adequately suffice, though readers should bear in mind that "science fiction" is often in reality "science fact".

Len Kasten states in *The Secret History of Extraterrestrials* that it is slowly dawning on us that the world we see around us is a cleverly designed illusion and that we have been expertly manipulated to believe it as reality. More and more spiritually-inclined people are questioning whether the "real world" is being "hidden from our view". The view that is gathering support is that we are manipulated by controllers who gain from our ignorance. But as Len Kasten says, the controllers have a large investment in keeping us from learning our true nature.

He comes from a slightly different thought process than Bramley, Icke and others in that he has a more positive take on alien presence: "It is beginning to appear that this may be the reason we are not being told about the extraterrestrial presence on Earth. If we were to be informed that we are now being visited by advanced aliens and that they have the ability to pull us out of our darkness and to shed light on our origins, and our spiritual potential, we would no longer settle for [being treated like sheep]".

From the previous chapters, this view seems somewhat optimistic.

Whether the aliens have a secret agenda of hybridisation; created a controlling elite; operate though a controlling elite; are part of the controlling elite; or here to help extract us from the controlling elite, - the extraterrestrial reality is something which we need to understand.

According to Michael Salla (referenced earlier), a leading writer and lecturer in the field of exopolitics (policies of governments, decision making and actions in relation to the extraterrestrial presence on Earth), there are four possible perspectives on the ET presence: firstly, the *intruder* perspective, that the aliens are here for their own purposes and intrude upon human life through their abduction and hybridization programs; secondly, the *manipulator* perspective, that aliens have been manipulating humans covertly since we first appeared on Earth; thirdly, the *helper* perspective, that aliens are here to help us grow in consciousness and solve our problems, including prevention of the use of destructive weapons; and, finally, the *watcher* perspective, that the aliens are here to observe the 'Earth experiment' and are agents of a larger galactic organisation who are powerful enough to 'step in' if necessary. The strongest evidence seems to support the idea of the intruder and manipulator perspectives, as can be seen throughout this book, but the helper perspective clearly also exists, from the experiences of the 'contactees' noted earlier with our "space brothers".

Elite control

The classic operating method of elite control involves a requirement that there is little regard or emotion relating to the mass of the population; they are there solely to carry out functions required by the elite rulers. This goes back to the time of the creation of today's human being, when the Annunaki needed a dispensable labour force to work the mines in Africa. To ensure control is absolute, their methods require the suppression of any individuality or spirit. And increasingly, the world envisaged by George Orwell in *1984*, his classic work on totalitarianism, is coming to pass. The knowledge of how to free oneself from mental and physical control has been systematically suppressed. Surveillance of every aspect of our lives by the State is complete and absolute. We must free ourselves from Orwell's vision of the future: "imagine a boot stamping on a human face – forever".

So how does the control system work? As David Icke says, when you are just a few people compared with the vast labour force you need to control, the critical feature must be centralised decision-making. You simply can't have decisions relating to policy on control and power taken anywhere other than at the top table. And the process of centralisation has now effectively covered all areas of human life.

Centralisation is achieved through what Icke calls the "pyramid of manipulation". He says: "The same compartmentalisation of knowledge used by the secret societies is repeated in the structure of government, banking, business and every other organisation and institution. Only a few at the top of the 'individual' pyramids know the real agenda and what the organisation is trying to achieve. The further you go down the pyramid, the more people are involved in the organisation, but the less they know about the real agenda. They are only aware of the job they do every day. They don't know how their contribution connects with those other employees in other areas of the company, government or whatever. They are 'compartmentalised' and the only people who know how it all fits together are at the top. Through the pyramid within pyramid structure, the elite families can coordinate the same policies through apparently unconnected, even 'opposing', areas of society. All roads lead eventually to them, - everything from the food we eat, the water we drink, the medical care we receive, the news we watch, hear and read, the entertainment we are given, the governments that dictate to us, the military who enforce the wills of the governments."

The process of control can be speeded up by what Icke calls the process of "Problem-Reaction-Solution." It is utilised to bring about changes that might otherwise be resisted by the mass of the population. Firstly one has to create a *problem* while blaming someone else, then create a public *reaction* of fear, outrage, and general demands from the people that 'something must be done', and then it allows a *solution* to be introduced by those in power which entails changes in society which restrict freedoms, but are deemed 'necessary' and accepted by the people for their safety. The more outrageous and traumatising the event that is utilised to create the problem, the easier it is for governments to take away our freedoms under the guise of our safety and security.

One man who certainly knew about the elite group that controls society is L. Fletcher Prouty, who from 1955 to 1963 was the liaison between the USA's Central Intelligence Agency (CIA) and the Department of Defence (DOD) for military support of what he calls "secret operations". As a result, in 1973 he wrote a book entitled *The Secret Team: The CIA and its Allies in Control of the United States and the World*. It is as relevant today as it was then, and states that there are secret operatives across politics, intelligence and military that control all sectors of government and society, right across the globe. Prouty says that the "Secret Team" are "the real power structure", and are unconstitutional and beyond governmental control. Their power has almost certainly increased over the years since Prouty wrote his book, and suggests that Icke's more contemporary analysis may not be far from reality.

The end goal is to produce, in every area of society, dependency on the elite rulers, as this is the way to ensure control. David Icke's work provides examples of mental, physical and emotional control across all areas of our lives, including chemical additives to food and water, electromagnetic pollution, manipulation of the DNA, mind control, information tracking and the microchip which in due course will be designed to exert direct control over the human body and its emotions.

Agenda 21 – the plan for world governance

On June 13, 1992, Agenda 21, an action plan proposed by the United Nations, was adopted by 178 governments around the world at the UN Conference on Environment and Development held in Rio de Janeiro. Agenda 21 is described as a "non-binding, voluntarily implemented action plan of the United Nations with regard to sustainable development".

One of the best explanations of what Agenda 21 really means to the future of the planet is Rosa Koire's 2011 book *Behind the Green Mask: UN Agenda 21*. In essence, Agenda 21 is about supplanting individual rights (as laid down in most of the constitutions of democratically elected governments) with central and, ultimately, global governance. Individual rights and freedoms are replaced with the 'rights of the community'. This is a 'legal' method by which elite controllers can remove individual rights surreptitiously without any serious objections being voiced.

It is a 21st Century ("21") action agenda for the UN, other multilateral organisations, and individual governments around the world that can be executed at local, national and global levels. Agenda 21 is now being implemented in many parts of the world, particularly in Europe and the USA.

It is a land use policy which the UN wants implemented in every city, county, state and nation. It should be borne in mind that the UN is much more than a peace-keeping force, although this is what it is largely known for. The idea is that individual rights in general are to give way to the needs of communities as determined by a global governing body, and land use decisions should be taken away from private property owners.

As Rosa Koire points out, the innocuous sounding aims of the policy cover the fact that the policy is a "whole life plan" involving the education system, the energy market, the transportation system, the government system, the health care system, food production and so on. It is a "plan to inventory and control all of the natural resources, means of production, and human beings of the world. The plan is to restrict your choices, limit your funds, narrow your freedoms, and take away your voice." It will allow far easier monitoring of humans. By urbanising populations, people become more and more dependent upon governments for all their needs. It appears to be one of many methods (see the earlier sections on control) by which humans can be made docile and controllable as required by a dystopian society.

Ms Koire, an experienced real estate appraiser, provides detailed examples in her book of how the plan is being implemented in the USA. She states that the implementation of the policy at local level is often assisted by the use of the Delphi Technique, which was developed during the cold war as a mind control technique. It is used to maneuver a group of people to accept a point of view that is imposed on them while convincing them that it was their own idea. The way mind control is utilised in society is addressed in the next section.

Agenda 21 has a specific goal called "social equity". This was set out originally in the 1976 UN Conference on Human Settlements which stated "private land ownership is a principal instrument of accumulation and concentration of wealth and therefore contributes to social injustice. Public control of land use is therefore indispensible." There is no desire to uplift the poor - quite the opposite. If a group of peoples, such as in Europe or the USA, have a high standard of living, the desire is to level those standards downwards. This is now well on the way to being achieved, as any average person in the western world will readily acknowledge. Living standards and spending power are being squeezed. This is assisted by unregulated immigration policies which lower standards of living, drain resources and push the cost of labour downwards. Indoctrination of obedience to the cause from the earliest possible age and reduction of free thinking is the way forward for today's youth. As Rosa Koire says: "Our culture is now conditioning us to be accustomed to the loss of privacy".

Ms Koire points out that the easy sell for the plan is the impending destruction of the environment (for more detail on this see the chapter on Planetary Destruction which follows). The sell is that we are "compromising the ability of future generations to meet their own needs". We are being urged to change the way we live as we are the cause of the planetary deterioration. The subplot is: Follow the global rules the controllers wish to implement and the planet can still be saved. Of course, this involves the removal of individual sovereignty and free will.

Mind control

The concept of mind control is a controversial subject. Control of the human mind goes against the concept of 'private thought', free will, freedom of thought and free speech. But this is what Controllers of society desire to do. The goal of government mind control is to control the human mind so that free will is eliminated and the mind responds favourably to external impulses.

The US military's interest in mind control goes back to the Korean War in the 1950's, where returning soldiers exhibited behavioural and personality changes as a result of capture, and what was termed "brainwashing" of these soldiers (a prolonged psychological process, designed to erase an individual's past beliefs and substitute new ones). As a result of these events, the US military began to study mind control through a number of classified programs.

In his outstanding 2006 book entitled *Controlling the Human Mind*, Dr Nick Begich describes various patents that have been issued as a result of research in the field of mind control and discusses how mind control occurs. He references an article published in Spring 1998 by the US Army War College entitled *"The Mind Has No Firewall"*, which stated that research had concluded that "the body is capable not only of being deceived, manipulated, or misinformed but also shut down or destroyed – just as in any other data-processing system." Technology appears to have now reached the point where emotions, thoughts, memory and thinking can be manipulated externally, including interference with short and long-term memory and the insertion or deletion of memories.

Nick Begich explains the process of how this technology can be utilised against the human mind, which makes it frighteningly apparent that we cannot physically block the technology from targeting any part of the body. In particular, electromagnetic fields can be introduced from devices outside the human body to target the brain or other parts of the body. Much of the development work for mind control was conducted through the CIA's extensive drug and mind control program in the 1960's, called MKULTRA, which had started out as Project Bluebird and then Project Artichoke.

Begich notes various patents in the field: a 1976 patent that relates to the monitoring of brain activity and then the alteration of it; patents in 2000 and 2003 concerningmanipulation of the nervous system by pulsing images displayed on a nearby computer monitor or tv; a 1989 patent concerningpain reduction using pulsed electromagnetic signals that cause ion flow in the nervous system; and a 1992 patent related to the use of very low or very high frequency audio signals for delivering subliminal information - a system that bypassed the conscious mind and dropped the information into the subconscious in a manner that avoided any conscious filtering of the information. In 1991 a method for changing brain waves to a desired frequency was patented. The brain's activity is mapped in order to read a person's emotional state, conceptual abilities and intellectual patterns. The brain's natural signal can be overridden causing the brain's energy patterns to shift. This is called "brain entrainment" which causes a shift in consciousness, and can also involve direct memory transfer technology. Finally, the extraordinary patent in 1996 entitled "Method and Associated Apparatus for Remotely Determining Information as to a Person's Emotional State", that relates to technology which can walk through any behaviour wall a person can erect and goes straight to the brain to see what might be on a person's mind.

As Begich points out, we have reached the point where the human mind and body can be controlled remotely without a trace of evidence being left behind. This has various obvious uses to governments both from a military viewpoint (such as the "Manchurian Candidate" scenario discussed below) and for civilian uses, a well-known example of which will be discussed below in relation to Cathy O'Brien's book *Trance: Formation of America*, subtitled *The true life story of a CIA Mind Control Slave*.

Robert Duncan's 2010 book *Project: Soul Catcher* takes the work of Dr Nick Begich further in relation to the art of bio-communication warfare. Duncan notes that the human being is a complex machine but the inner workings of human beings have now been deciphered. In particular, he notes that the mind has no firewall and no anti-virus software so the public is very vulnerable to psychotronic attacks. Little defensive research work appears to have been undertaken in the field of trip wires to mind hacking, only offensive work. The array of examples given by Robert Duncan of psychotronic viruses and other methods for influencing the psyche are enormous, and show the extent of the danger to humans.

This includes "voice of god" direct neural linking technology (voice transformation trickery/voice-to-skull technology) and brain entrainment which eliminates free will.

Controlling the human mind is an important part of control of human populations. Mind control can be directed against large population groups or against specific individuals, depending on the form of mind control utilised.

In relation to large population groups, devices such as HAARP are most effective for mind altering effects. The HAARP device, which was referred to in the exopolitics chapter in relation to the creation of a global defensive electromagnetic shield, and is referenced below in the chapter on Planetary Destruction in relation to weather modification, also has uses related to control of society and specific targeted populations. Predominant brain waves can be driven or pushed into new frequency patterns by external stimulation, utilising VLF and ELF waves, causing changes in thoughts, emotions and physical condition, which could be utilised to affect large population areas. Research has found that the brain can easily be entrained or influenced to change states by external electromagnetic fields. HAARP has the potential, therefore, to be used as a device for, amongst other things, control over populations or individuals that the elite controllers may consider dangerous. In the wrong hands, such as an elite group desiring to control society, HAARP can help controllers or governors of people become an oppressor rather than something which benefits the masses.

In relation to individuals, mind control methods have been extensively documented. The public has been introduced to the reality of government-sponsored mind control through blockbuster films such as the *Bourne trilogy*, as well as private use of mind control/memory alteration through films such as *Eternal Sunshine of the Spotless Mind, Wanted,* and *Trance*, amongst many others. Perhaps the most controversial film on the subject, however, is *The Manchurian Candidate*, which is based on real life concepts of government-sponsored mind control of individuals.

According to Colin Ross, MD, a "Manchurian Candidate" is an "artificially created multiple personality, where there's another identity inside which is given the mission parameters and carries out the mission. The idea is that if they are caught and interrogated, the front person has no memory of the mission; so it keeps the information secure." Ross goes on to state that the Manchurian Candidate is an experimentally created dissociative identity disorder (formerly known as a multiple personality disorder) that has been created deliberately, with a new identity implanted, with amnesia barriers created, and used in simulated or actual operations.

The name *Manchurian Candidate* derives from a book written in 1959 by Richard Condon in which a group of American POWs in the Korean War are brainwashed while crossing through Manchuria to freedom. They arrive back in the USA unable to recall the period of brainwashing, and one of them has been programmed to be an assassin. His target is a candidate for the Presidency of the USA. He is controlled by his Asian handlers through the use of a hypnotically implanted trigger, which they can utilise when the time is right.

From the beginning of the 1950's, the US Government agency responsible worked extensively on the creation of amnesia barriers, new identities and hypnotically implanted codes and triggers. Experiments were successfully undertaken in which people could successfully pass from the fully awake state to a deep controlled hypnotic state via the telephone, or some subtle signal, code or word that could not be detected by any other person present at the time. These signals, codes and words could be used at any time, in public or private. Control of the person hypnotised could be passed from one person to another without difficulty. The person believed their created identity so completely that they could pass a polygraph test. This allowed relatively risk-free use of such persons as spies and couriers of information.

There are many methods of creating dissociative identities, ranging from use of drugs, sleep deprivation, prolonged psychological isolation, electroconvulsive shock (sometimes known as depatterning and psychic driving). Dr Eastabrooks, a leading mind control doctor who wrote books such as *Hypnotism* (1943, revised 1957), stated that the key to creating an effective spy or assassin rests in splitting a man's personality, or creating multipersonality, with the aid of hypnotism. This view was reiterated by the leading Russian psychiatrist, Dr Smirnov, who has been referred to as the father of psychotropic weapons in Russia. Eastabrooks discussed fracturing personalities through induced trauma and then reprogramming people afterwards. The brain can contain a large number of fragments of thoughts, behaviours and personalities.

Cathy O'Brien wrote a harrowing account of her use as a "CIA mind control slave" in her 1995 book *Trance: Formation of America*. The book contains specific names and details, but these will not be repeated here. The principles behind her mind control are more relevant, describing how individuals can be utilised against their free will to assist a ruling group of persons to control society and achieve desired goals. Her mind was systematically fragmented so that her handlers could multi-task her, ranging from the use of sexual favours to assist deal finalisation, to acting as an information courier, amongst many other tasks.

Mark Phillips, who assisted in her de-programming work, wrote an introduction in which he stated that the programming that Ms O'Brien had undergone was known as trauma-based mind control, which is absolute, rather than other forms of mind control such as chemical or electronic manipulation, which are considered by experts to be only temporary. At the root of mind control is the principle that control of *human suggestibility* is recognised as the fundamental building block for external control of the mind. Global corporations understand this well, and the way products and services are advertised can be effective for mind manipulation and behaviour modification.

Cathy O'Brien was a perfect government target for mind control due to her traumatic early childhood abuse experiences, and so she was recruited into a government sponsored mind control project called Project Monarch, as she already had Multiple Personality Disorder (MPD). MPD is the mind's way, its defence, to dealing with trauma that is too difficult for the mind to assimilate. The memory creates a compartment that shuts itself away from the rest of the mind, allowing the rest of the mind to function 'normally' as though nothing had happened. The compartmentalization of the brain is created by the brain shutting down neuron pathways to a specific part of the brain, and conditioning the mind to similar abuse when it reoccurs.

Effective mind programming of such a person will shut down their emotional responses, remove free will, keep them financially dependent, and ensure that the various compartments of the brain can be accessed by handlers as required for covert activities or further abuse without any significant risk of detection.

Control of society through debt creation

Global control ultimately requires Super States to be formed, and this is well under way with the blocks of countries now forming into regional unions. But the creation and maintenance of an 'empire' is vastly different today from the days when the great Roman and British Empires were in their ascendancy. Then, the creation and maintenance of the Empire, and the plundering of the resources of others, depended upon military power or the threat of it. Moreover, in those days, the Empire actually 'ruled' over vast tranches of the planet.

Now, it is different. 'Empire' is a more subtle creature. Only in extreme cases does imperialism actually involve military muscle. Countries are generally dealt with differently from the methods used to control people, and the primary control method is now through a much simpler method than war: debt creation. This concept of modern control has never been better explained than by John Perkins in his classic book *Confessions of an Economic Hit Man*. This method is so subtle that most of the citizens of these 'independent' countries around the world don't actually realise that power lies beyond their own government and jurisdiction.

Since the devastation of Europe (including Russia) as a result of the Second World War, there has been a relentless drive to create a truly global empire, and it can realistically be said that America has now achieved that aim(although all empires collapse at some point too). For now, the USA rules over an almost perfect 'global empire', the like of which has never been seen before.

The concept of debt creation was firmly established after the Second World War as the means to control the world. This was achieved through the establishment of the World Bank (comprising the International Bank for Reconstruction and Development and the International Development Association) and the International Monetary Fund, both of which were largely funded by the USA and other Western nations. This allowed the United States drive for global domination to take shape quickly. The USA also utilised the United States Agency for International Development (USAID), and through these financial lending sources, the USA ensured that loans were 'tied' to the use of US construction and engineering companies. The end result was that whilst American corporations became rich, third world countries rapidly became poverty stricken, weighed down by an impossible debt burden.

The concept of using debt to plunder resources and control nations was so brilliantly conceived, that global domination was 'institutionalised'. The financiers, economists, developers and lawyers relentlessly fanned out across the world to close out infrastructure and other development projects through international loans. Government itself was not the main architect of the global empire; banks and major corporations were.

But what happens when the borrowing country defaults because of the overwhelming debt burden? The lenders demand their "pound of flesh". This may involve control over UN votes, installation of military bases, or access to precious resources which can be developed at great profit by preferred global corporations.

Rarely does it become necessary to go further - into military action. This is because debt creation has in itself allowed almost complete control of most nations.

Unfortunately, planetary control is leading to planetary destruction, and the likely end of our world as we know it, as will be shown in the next chapter.

Planetary Destruction

The short sighted, self-serving attitude of today's civilisation with its concentration on greed and profit is quickly bringing the planet to a point of environmental disaster, as can be seen from the reviews below.

Remote viewing of environmental disaster – Courtney Brown

An unusual source depicting the coming environmental disaster is provided by the remote viewer Courtney Brown PhD in his book *Cosmic Voyage*. His research indicates that the planet is destined to undergo an environmental disaster which will render the surface of the Earth uninhabitable for centuries, causing remaining inhabitants to live below the surface or in domes.

He was able to conclude this information through remote viewing sessions he undertook on the subject. Using scientific remote viewing (SRV), the information coming from the unconscious is recorded before the conscious mind has a chance to interfere with it through rationalization or imagination. The process of 'target' remote viewing has been utilized by the US military, and Brown was trained using the very strict SRV protocols which blocks out activity of the conscious mind.

As was shown in the chapter on contactees and abductees, contact with extraterrestrials often includes a warning about the wanton abuse and destruction of the environment of Earth. Brown had read John Mack's *Abduction* in which environmental abuse by humans was a central feature, and of major concern to ET's, and so Brown decided to remote view the near future of Earth under the guidance of a trained monitor. He moved to viewing Earth in about the year 2300, and watched a polluted city in which there were too many people for the system to handle, resulting in health problems, sickness, disease, food and housing shortages. The ecosystem had completely broken down due to usage of traditional energy sources, and there was complete disregard by humans for life other than their own. The elite already lived underground in compounds away from the masses.

Brown understood from the remote viewing that humans would need the help of the Martian and Grey extraterrestrial races who had long experience of living underground.

It is apparent at this time that there are too few people on Earth concerned about the failing environment, and there is little they can do anyway due to the global lack of concern. This is likely to lead to humans having to endure immense hardship as the price of learning how to respect life other than their own.

At a further remote viewing session on the future culture of Earth, Brown saw that the Earth environment degenerated over a period of several hundred years from the current time, resulting in apocalyptic scenarios with roaming gangs and a focus on special sanctuary-type human habitats. Extraterrestrial leaders will not assist, as humans must get themselves out of the environmental disaster. No bailout is possible - not because galactic leaders cannot do that, but because humans must not become a dependent species, but rather a mature and helpful species.

The environmental crash comes from chasing physical happiness exclusively at the expense of the planet's carrying capacity, inevitably destroying most of the human population. Brown's remote viewing concludes that the required maturity only comes through experience. Learning comes through experience. Fortunately, Brown is able to conclude that there will be a rebirth of humanity after a lengthy period of great hardship. This process will be assisted by the Greys and Martians, both races being here on earth already.

Linda Moulton Howe's *Glimpses of Other Realities* also contains clear warnings of impending environmental disaster on Earth. In an interview with abductee Jim Sparks, Sparks revealed that during his interaction with reptilian alien creatures, the aliens stated that they had had contact with the US Government but their warnings had gone unheeded. The aliens stated that Earth's air and water are contaminated. The forests, jungle, trees and plants are dying. There are breaks in the food chain. The planet is overpopulated. The governments had been made aware of technology to rectify the situation but feared losing their power base by doing so. Now the aliens have reached the point where they are worrying that the investment they have made in Planet Earth is at risk so they are taking measures to ensure that life on Earth can continue even if humans do not change their current direction and destroy the environment. The aliens insurance policy is that they are collecting seeds from plants, animals and humans. Through semen and ova extraction, the aliens can restart humanity or other earth life all over again either on Planet Earth or somewhere else.

In a separate interview with Linda Porter, Linda Moulton Howe was told that through Linda Porter's contact with aliens she had been told that the aliens have discovered that the soil, vegetation and water are now contaminated through a chemical poison that derives from testing that humans have done in outer space. Betty Andreasson Luca similarly confirmed through her alien encounters that the human race will become sterile by the pollution that is on Earth, and the aliens are taking seeds so that the human form will not be lost. Linda Moulton Howe's last comments in the Epilogue to her extraordinary book is that perhaps the non-human intelligences which watch Planet Earth see the future of the human species as hanging in the balance and will not allow humans to destroy the earth. She suggests that perhaps this is why we are seeing the increased presence of these beings in human affairs despite aggressive government efforts to keep them under wraps.

Weather manipulation

One of the biggest threats to the continued survival of humanity is the HAARP device referenced in the chapter on Exopolitics. Some of the first researchers on the effects of this device were Jeane Manning and Nick Begich who, in 1995, published a shocking major investigation into HAARP entitled *Angels Don't Play This HAARP: Advances in Tesla Technology.*

They revealed that the original patent for HAARP stated that the idea of the invention is to generate a beam of radio waves of enormous intensity and direct this toward the upper atmosphere. At certain altitudes, electron cyclotron resonance heating of existing electrons would cause further ionization of the neutral particles of the atmosphere. Among the intended uses of the invention are to 'disrupt microwave transmissions of satellites', or cause 'even total disruption of communications over a very large portion of the Earth'. Other intended uses include weather modification, lifting large regions of the atmosphere, and intercepting incoming missiles.

Earth is protected from the radiation from the sun (gamma rays, X-rays, and shorter wavelengths of ultraviolet) by protective layers. When the rays hit the outer layers of Earth's atmosphere, they are absorbed by atoms, which releases electrons and changes the atoms to positively charged ions. The ionosphere stretches from about 30 miles above earth to around 300 miles up. This natural electrically charged shield around the Earth filters harmful wavelengths of solar radiation, protecting the Earth from significant bombardment. Above that layer is the Van Allen radiation belt at between 2,000 and 12,000 miles high, and below the ionosphere is the ozone layer. The ionosphere creates the electrojet, a jet of electrically charged particles which follow Earth's magnetic field lines. The electrojet flows in the form of a direct current into the polar regions of the Earth. HAARP enables changes to occur in this electrojet. But as Manning and Begich's book asks: Who would want to change the electrojet? Who would want to control the weather? The answer is, of course, the US military.

The US military has a history of weather-related experiments. They exploded three atom bombs in the Van Allen radiation belt in 1958, which produced artificial belts similar to the Van Allen natural radiation belts, and shot a curtain of radiation around the world. Then in the 1960's they decided that the ionosphere ought to be controlled because it was unpredictable. They therefore attempted to replace a ten by forty kilometer section of the ionosphere with a "telecommunications shield" of 350,000 copper needles tossed into orbit. Unfortunately it seems that the band of copper wires interfered with the planetary magnetic field, so further activity related to this program was curtailed. It is, therefore, not a complete surprise that the US military would wish to utilise a device such as HAARP to control weather, amongst many other uses.

In terms of weather adjustment, HAARP has the ability to move the upper-atmosphere jet stream, changing global weather to one country's advantage, and also aiding military defence capability. The power of HAARP can literally pierce the natural shield which protects us all from being bombarded, ultimately to extinction, by cosmic radiation. The slicing of the ionosphere and the ozone layer may already have occurred. This would potentially allow harmful radiation to penetrate Earth, if the information contained in a lecture in January 2014 by Dane Wigington of GeoEngineeringWatch entitled *Climate Engineering Weather Warfare and the Collapse of Civilisation* can be believed. The lecture/video shows dramatic changes in the weather patterns and the use of substantial aerosol spraying from aircraft. Perhaps the spraying is an attempt to compensate for the rips in the protective layers above Earth.

The spraying, according to his lecture, is certainly causing significant damage itself. His lecture, on the website GeoEngineeringWatch.org, paints a frightening picture of rapid planetary deterioration, and can be found at: http://youtu.be/5yZhh2leRJA

Planetary Damage – Thom Hartmann

It is a reasonable view that the process of control of people and nations has not just damaged humanity, it has damaged the planet itself, as was so clearly portrayed by Thom Hartmann in his book *The Last Hours of Ancient Sunlight*. Hartmann's book was written about a decade and a half ago, in 1999, but clearly shows how we (citizens of Earth) are getting close to wiping ourselves out by our drive for profit and greed as we deplete critical resources required for our survival (such as trees for oxygen, drinking water and so on). Hartmann's book can be seen as a "wake up call" to the world. Indeed Neale Donald Walsch said as much in his "afterword" to the book: "When I first read this book I knew I could never view my life in the same way again. I could see myself as part of the problem, or part of the solution, but I could never again see myself as having nothing to do with either." It is about waking up to the complete destruction of the planet. There seems to be a complete lack of awareness, lack of caring through wanton greed and thinking about "now", the service of one's needs today only. We need to make huge changes in the way we see and understand the world and recognise the need to control our population, save our forests, reduce wasteful consumption and re-create community. Key features discussed by Hartmann are overpopulation, depletion of resources, pollution and toxic air, the combined effects of which cause planetary destruction.

Overpopulation and depletion of resources

Global control by the ruling elite is assisted by poverty and dependency. Overpopulation, which increases by 1 billion every decade now, is not helping planetary survival although it does help planetary control by the ruling elite. The figures for population expansion are shocking. In 1800 there were 1 billion people on the planet, in 1930 this increased to 2 billion, by 1960 this hit 3 billion. Thereafter, we have added one more billion in each further decade. We are currently adding more people at around the size of a city like Los Angeles about every 2-3 weeks. The reason the population can increase so easily is that we are sustained by oil and gas, the "ancient sunlight resource", and this can't continue. This resource is depleting rapidly, and it is expected that by the time we hit 10 billion people there will be food for only about 3 billion. This will create famine, plague, and war amongst other things.

Once the resource of oil, gas and coal is depleted, we will go back to the numbers that the world could support before their discovery: about half a billion. Our so-called "advanced modern culture" has destroyed much of our world, perhaps to the point already where we will see the end for most people within our children's lifetime. Fossil fuels have essentially powered the expansion of our population from half a billion before their discovery to over 7 billion today. And when these sources are depleted, as they will be, the population will reduce. "Growth" is seen as "good"; yet, it is straining nations to breaking point as they can't support the "growth". And if one is not part of the ruling elite, you won't have the wealth and power to survive when times are bad. When resources get sparse, bad things happen: disease, war, starvation, horrible death. The citizens of the richer Western nations believe they are immune from the world's hot spots, the horrors of Haiti, parts of Africa, and so on, but in the end it gets us all.

As Richard Preston shows in his 1994 book *The Hot Zone*, one of the most frightening books ever written, the decimation of the rain forests of Africa has unleashed on the world a wave of deadly viruses on its human invaders, such as HIV, Marburg and Ebola. And greed and profit was at the heart of the invasion of the rain forests. Preston states that the emerging viruses are surfacing from ecologically damaged parts of the earth. The earth is beginning to react to the "human parasite, the flooding infection of people, the dead spots of concrete all over the planet, the cancerous rot outs in Europe, Japan and the United States, thick with replicating primates, the colonies enlarging and spreading and threatening to shock the biosphere with mass extinctions." Preston goes on to state that nature has interesting ways of balancing itself. The rain forests (where the viruses emerge from) have their own defences. Preston says "The earth's immune system is seeing the presence of the human species and is starting to kick in. The earth is attempting to rid itself of an infection by the human parasite". Brilliant, damning, language from a master writer.

Chemical pollution into water and soil are further hazards which will decimate us. Desert zones and infertile lands are increasing. Seas have dead zones now. The end appears to be nigh. And who loses and who wins from the greed and destruction? As Hartmann says "Around the world we find that rapid growth is straining virtually all nations, and the greatest pain is usually experienced by the individual people and families who do not share the extreme power and wealth of the society's ruling elite (whether the elite is corporate, governmental or military)".

Toxic air

That's if we don't die of the toxic air. The rainforests are being cut down at a fast rate, and they largely provide the oxygen we breathe to keep us alive. Large areas of rainforests are being cut down to satisfy corporate greed and provide cheap grazing for cattle that are butchered and exported to provide beef for cheap burgers in the West. Water is also becoming undrinkable in many parts of the world due to contamination of groundwater. Empires decline when they can't get hold of resources; war becomes inevitable in these circumstances.

We can take the view that we are helpless to stop the inevitable end and escape from the harsh reality into mindless television, drug addiction and the like, or we can (as Hartmann recommends) take steps such as downsizing, moving to the country and becoming self-sustaining, accepting the concept of "enough", sharing, living in community (as in creating happiness through being with people), passing the word around about waste and destruction and relying on the "100 monkeys syndrome" where change happens as more and more people live different lifestyles.

Perhaps we should listen to the words of Al Gore in his latest book *The Future*. He wants people to "get involved" and commit to actions which help a "sustainable future" for Earth. If we all do our own little bit for the environment, the ripple effect will then do the rest. Unfortunately, most minds are far removed from the point where they can think beyond themselves and their own needs. But one has to hope that things will change. Even Gore, a wealthy and influential politician, thinks that the super-wealthy elite have rigged the financial system, are careless with the eco-system, have corrupted the political system, and bought the media.

The ravaged landscape

In Nick Redfern's book *The NASA Conspiracies*, he recounted a story of a scientist in a laboratory who had been examining an alien body which had a one-piece fitted outfit on it, that extended from the lower neck right down to the bottom of the feet. To remove the outfit took several hours and required cutting the material. The scientists analysing the body discovered that the suit was "alive" or had a built-in memory, because not only would the fibers bond when brought close together, but they also appeared to bond with the exact same corresponding fibers time and time again. But when the scientist tried on the suit himself, he began to feel claustrophobic and started to receive disturbing images in his mind of a dark and frightening future for the Earth and for all life on the planet, and particularly for the human race. Nick stated that the future depicted was of an irradiated world, ruined cities, huge atomic mushroom clouds looming miles upward into an ever-black sky, as strange objects resembling flying saucers flew across the ravaged landscape. The human race had been reduced to absolutely minimal levels.

Will things change before it's too late?

PART 6

Outside Influences
and Signs

They Left Signs That They Were Here

If you knew cataclysms had occurred in the past, and a further cataclysm was coming, through your understanding of astronomical alignments and movements, but were not exactly sure when such a disaster would befall civilisation, would you want to say to future survivors that you were here before them? Or leave a road map to show what earth looked like before the changes occurred? If so, how would you do it? It seems that now-lost civilisations from the past did leave signs: The so-called "sacred sites". Incredible structures that we can still see today, even if today's archaeologists dismiss such suggestions. It takes a bit of imagination, and a desire to believe, but the evidence is there for all to see. Moreover, the "sacred sites" across the world were purposefully arranged.

After the occurrence of the cataclysm, you would somehow want future civilisations to know that you were technologically advanced, and somehow to try and retain scientific, agricultural and the many other areas of know-how and knowledge as far as possible in the survivors. When the cataclysm arrived, the survivors would span out around the globe seeking safe places to live and survive until the planet had stabilised again.

They may already have picked the places where survival was most likely.

The Cataclysm and Antarctica

One such cataclysm took place around 17,000 years ago when according to the theory of Professor Charles Hapgood, who was Professor of Geology at the University of New Hampshire, earth-crust displacement occurred. This shifted the temperate area of Antarctica 30 degrees of latitude, or around 2,000 miles further south, out of the warm climate it existed in, and into a much colder climate. This saw Antarctica become increasingly buried under hundreds of thousands of tons of ice, and the loss of whatever existed on that land before it shifted. At its peak 17,000 years ago, most of Europe and North America was buried under ice two miles thick, and as the thaw occurred, the world's seas and oceans rose by over 400 feet. It is not clear why the pole shift occurred when it did, or why the ice age ended when it did (Paul LaViolette proposed in his 1997 book *Earth Under Fire* that a galactic superwave melted the glaciers), but it seems that certain highly intelligent inhabitants of Earth at that time had understood that a change in Earth was coming and they wanted to map how Earth looked before the cataclysm occurred so that there would be markers to the past afterwards. Perhaps with their flying crafts, the inhabitants could see that the massive ice blocks were breaking up and that, once this occurred, Earth would be instantly devastated by a massive surge of flooding.

Hapgood's theory derives from the discovery of a genuine map, known as the Piri Reis Map, which was made at Constantinople in AD 1513. This map was drawn by Admiral Piri Reis and shows an ice-free Antarctica. Clearly Piri Reis did not acquire his information from contemporary explorers, because Antarctica remained undiscovered until 1818. The puzzle is that we know that Antarctica was swallowed by ice well over 6,000 years ago. Piri Reis, however, makes it clear that he was merely a compiler and copier, and his map derives from older source maps dating back to the fourth century BC or earlier.

Hapgood suggests that the source maps that Piri Reis utilised had themselves been based on older source maps going back into time unknown. Hapgood asserted that this was irrefutable evidence that the earth had been mapped in detail before 4000 BC by an unknown civilisation which had achieved a high level of technological advancement, voyaged from pole to pole, and explored Antarctica when its coasts were free of ice. Incredibly, whoever these peoples were that had mapped the earth, they had an instrument of navigation for accurately determining longitudes that was far superior to anything possessed by the peoples of ancient, medieval or modern times until the second half of the eighteenth century.

Fascinated by the Piri Reis map, in 1959, Hapgood examined further maps held by the Library of Congress, and was astonished to find another map, the Oronteus Finnaeus map from 1531, showing Antarctica much as it looks today on a modern map, but without ice. This makes clear that Antarctica was definitely mapped more than 6,000 years ago. The Piri Reis map also shows West Africa as having an ample water supply, such as lakes that no longer exist in the Sahara. Between

10,000 and 6,000 years ago, the mistral, the north wind, was very wet, carrying moisture from the melting glaciers of the Ice Age so that the Sahara was green and fertile. Since the Piri Reis map shows West Africa with lakes, it would seem that the original map used by Piri Reis dates from that period.

It is, therefore, a reasonable assumption that a human civilisation had mapped Antarctica before the earth-crust displacement over 17,000 years ago and had subsequently disappeared, or its survivors had moved on and lived as best they could, retaining knowledge for a future civilisation which would come again once the Earth stabilised. The Piri Reis map shows that the survivors would have known other parts of the globe because the western coast of Africa, the eastern coast of South America and the northern coast of Antarctica are clearly depicted.

Earth is an inherently unstable place. Indeed, in 1955, Hapgood noted from the study of rocks that since the beginning of geological history, the geographical poles have shifted their position as many as two hundred times, including sixteen shifts during the Pleistocene era alone. Geologists understand that pole shifts occur quite regularly. These shifts change the earth's axis relative to its surface, leading to the disappearance of the old polar ice-caps and the appearance of new ones in areas that had previously experienced more temperate climates.

Hapgood concluded that crustal displacement had almost certainly taken place on at least three occasions during the past 100,000 years, causing tremendous periods of upheaval over probably several thousand years on each occasion of a pole shift. Hapgood concluded from the available evidence on the Pleistocene epoch that, from around 50,000 years to around 9,600 BC, the northern polar cap had been located somewhere in the area of Hudson Bay in Canada, meaning that large areas of North America would have experienced ice sheets of two to three mile thickness. Then, as mentioned above, at around 9,600 BC, a shift of the earth's crust took place, which Hapgood was unable to explain. It saw North America 'slide' southwards, taking with it the whole western hemisphere, while on the other side of the globe the eastern hemisphere was equally tilted northwards. This crustal displacement would have taken around five thousand years to complete, and certainly corresponds to the climate changes which occurred in the Egyptian/Mesopotamian regions about 10,000 BC. The actual polar shift would have occurred in a blink of an eye. Russian scientist Immanuel Velikovsky concluded this must have been the case from investigations into the Beresovka mammoth, found frozen in Siberia in 1901 in a half-standing position with buttercups in its mouth, still edible.

Add to this the onslaught caused by the end of the Ice Age - from its glacial maximum 17,000 years ago to its stabilisation 10,000 years later, and it is not difficult to see that survival on earth would have been a monumental struggle. No one knows why the Ice Age suddenly ended, but huge areas of coastal land were submerged under the ensuing floods, as much as 10 million square kilometres of land.

It is a certainty that earth-crust displacement will occur again, and indications are available to us through modern technology to show that this process has already started, even if actual crust movement may still be many years away. This is because earth life runs in cycles, and there is nothing humans can do about that. When it happens, as we can see from the Velikovsky investigations, it will happen very fast, and then take many subsequent years to stabilise. This scenario is depicted in the 2004 film *The Day After Tomorrow*. What will we leave behind for future civilisations?

The Sacred Sites and Atlantis

An extraordinary book entitled *The Atlantis Blueprint*, by Rand Flem-Ath and Colin Wilson, utilised the data researched by Professor Hapgood in his books *Earth's Shifting Crust* and *Maps of the Ancient Sea Kings* and went much further. It reveals that there is a single global pattern that ties the sacred sites across the world from Stonehenge and Avebury in the UK to the incredible pyramids, stones and monuments in Egypt, China and South America. This implies the existence of an advanced civilisation that existed before the flood and earth changes and managed to communicate important geodesic, geological and geometric information to people who became ancient mariners and re-charted the globe.

The book propounds the theory that, sometime before the devastating earth crust displacement, scientists of that time recognised the increasing earthquakes and rising ocean level were a warning of a coming geological catastrophe. So geologists decided to try and determine how far the crust had shifted from the last earth-crust displacement so that they might get some idea of what they might expect to face in the future. Consequently, they left geodetic "markers" at the places that they considered crucial to their calculations. These sites have been designated as sacred but, as *The Atlantis Blueprint* indicates, the sites actually had a scientific purpose. Of course, the "markers" were awesome, and the construction feats of the monuments erected by the now-lost civilisation are a wonder to behold. Much of these are difficult for us to replicate even today.

As the title of the book suggests, before Antarctica was covered in ice following the last polar shift, the land mass was nothing other than the fabled lost civilisation of Atlantis, where technologically advanced inhabitants lived.

It seems that the Giza Pyramids were built on the centre of the earth's land mass for a very specific purpose. Calculating outwards from the Pyramids, utilising the "Golden Section" (phi) system, all the sacred sites of the world lie on specific latitudes aligned with either the old or new positions of the North Pole. These sites are systematically listed in *The Atlantis Blueprint's* Appendix. They are a list of, and specific positioning of, all the famous sacred sites of the world. It seems that people from the past had an advanced knowledge of the earth's dimensions, and used a sacred geometry system to measure distance and positioning, when modern society has been led to believe that no such civilisation or knowledge could have possibly existed.

Colin Wilson gives a helpful insight in *The Atlantis Blueprint*. He says our long gone ancestors were far more intelligent than we give them credit for. He says they had an intelligence that was different to today's leaders of civilisation. Their intelligence was intuitive, a right-brain intelligence rather than today's analytical left-brain intelligence. Wilson says the intelligence of modern man is like a microscope to early man's telescope intelligence. Modern man *narrows* his senses to study minutiae, but Wilson says our ancestors *widened* their perceptions to try to understand the cosmos. It seems that modern man has lost the wider dimension in focusing on minutiae.

The incredible thing is that what becomes apparent from a reading of *The Atlantis Blueprint* is that over *sixty* sacred sites relate directly in terms of alignment to the Hudson Bay Pole, which moved in 9,600 BC, the most well-known sites from these sixty sites which could still potentially be visited include Baalbek, Canterbury, Chichen Itza, Cuzco, Easter Island, Giza, Jerusalem, Lhasa, Luxor, Machu Picchu, Nasca, Newgrange, Quito, and Rosslyn, and eight relate to both the Hudson Bay Pole and the Yukon Pole which moved in about 50,000 BC. These eight sites are Byblos, Jericho, Nazca, Cuzco, Xi'an, Aguni, Pyongyang, and Rosslyn Chapel. Incredibly, it gets even better. There are a further three unique sites, Avebury, Abydos and Nippur, which are one degree out when aligned with the both the Yukon Pole and Hudson Bay Pole. It seems that sometime between 50,000 BC and 9,600 BC, a cosmic event had caused a one-degree shift in the earth's crust. This is about a 70-mile shift, rather than the 2,000 mile shift in 9,600 BC. This would explain the reason why these special sacred sites are misaligned by one degree. As Rand Flem-Ath surmises, it could have been the "sharp shock warning" which pressed the "Atlantean" civilisation into placing geodetic "markers" at the positions that they considered crucial to their calculations for a future cataclysm.

Svalbard Seed Vault

All civilisations want to survive cataclysms. If they have the ability to prepare for future-forecasted cataclysms, they will. Some say another cataclysm on earth is due, and our modern governments appear wisely to be preparing for it.

The Svalbard Global Seed Vault was officially opened on 26 February 2008. This vault, which has begun storage of seeds, is situated in a Norwegian archipelago in the Arctic Ocean, about midway between continental Norway and the North Pole, and is reported to have the capacity to store 4.5 million different seed samples. Since each sample will contain on average 500 seeds, around 2.25 billion seeds may be stored there. They are being held in the Arctic storage facility in case a future disaster wipes out food crops.

The seeds in the Svalbard Seed Vault are duplicate samples, or 'spare' copies, of seeds held in genebanks worldwide. This vault will provide insurance against the loss of seeds in genebanks, as well as a refuge for seeds in the case of large-scale regional or global crises. The location was carefully chosen to provide maximum protection to the seeds.

The underground vault has been built as a 120-meters long tunnel inside a mountain, at about 130 meters above sea level on Spitsbergen Island. Permafrost and the thick rock will ensure that the samples remain frozen, even without electricity, and the site remain dry even in the event of the melting of icecaps. Seeds are packaged in special four-ply packets and heat sealed to exclude moisture.

Spitsbergen was considered ideal due to its lack of tectonic activity and its permafrost, which will aid preservation. Prior to construction, a feasibility study determined that the vault could preserve seeds from most major food crops for hundreds of years. Some seeds, including those of important grains, could survive far longer, possibly thousands of years. Preparation for the next cataclysm is well underway.

The Monuments

As mentioned previously, the civilisations of time gone by left monuments to remind us that they existed, such as the Pyramids of Giza in Egypt and the city of Tiahuanaco in Bolivia. As Philip Coppens points out in *The Lost Civilisation Enigma*, civilisation is far older than we assume from what archaeologists lead us to believe. Civilisation existed in Europe, Africa and the Middle East long before the so-called start of civilisation in Mesopotamia around 4,000 BC.

Andrew Collins cites in *Gods of Eden* the example of Nevali Cori, constructed over 10,000 years ago on the upper Euphrates of Eastern Anatolia. Nevali Cori can hold itself out to be a more realistic description as a "cradle of civilisation" than ancient Mesopotamia, which emerged a full 5,000 years later, and which may have been an area which the Aryans lived at one time. This was a culture sufficiently sophisticated that it was able to produce carved stone pillars so beautiful that they were more in keeping with the megalithic art of Malta or Western Europe, which were created many thousands of years later. The art had immense similarities to the Kalasasaya palace court at Tiahuanaco in Bolivia, a mysterious city which is referenced shortly.

Coppens cites many examples across the globe, but his description of Gobekli Tepe, the region where Andrew Collins' Aryans may have lived (which he describes in his book *From the Ashes of Angels: The Legacy of a Fallen Race*), in the highlands of Turkey near the Iraqi and Syrian borders, is particularly fascinating. The place consists of a series of mainly circular and oval structures set in the slopes of a hill. Here we can see that people had built extraordinary towns and structures as long ago as 10,000 BC. Unlike the discovery of Jerico, the town which dates back to 8,000 BC and created immense interest because of its biblical links, an older discovery in southeastern Turkey cannot raise huge media interest. Gobekli Tepe was not alone; another site, Karahan Tepe in the Tektek mountains, has been dated to 9,500 BC and contains carvings similar to those at Gobekli Tepe.

It is from Graham Hancock's immense research in *Fingerprints of the Gods*, however, that we can really get an idea of what was left behind for future civilisations to see. The problem is that because of the devastation of earth in the period from 17,000 BC to 7,000 BC, apart from the huge megalithic structures that still remain, most of what could have helped archaeologists has been washed away, buried or destroyed. But what is left still stretches the imagination. Hancock goes into great detail about why, in defiance of the viewpoint of conventional archaeologists, sites like Tiahuanaco may well go back 12,000 years or more. Orthodox archaeology concentrates on the latest layers of occupation and construction without considering the possibility that the origins of the site might be much older. It is well accepted that sacred places remain sacred forever, and new religious constructions, such as temples, are often built over the site of older temples or buildings. Teotihuacan in Mexico for example is highly likely to have been a holy place for thousands of years before 4,000 BC, and Mayan script tablets dating to over 12,000 years old have been found in the region. As Zecharia Sitchin points out in *When Time Began*, Sumerian, and then Akkadian, Babylonian and Assyrian kings, recorded in their inscriptions, with great pride, how they repaired, embellished, or rebuilt the sacred temples and their precincts. In the early twentieth century, a joint expedition of the University of Pennsylvania and the Oriental Institute of the University of Chicago spent many years working to unearth the Temple of Enlil in Nippur's sacred precinct. The excavators found five successive constructions between 2200 BC and 600 BC. The report noted that the five temples were "built one above the other on exactly the same plan". This principle applied just as precisely in the discovery of later temples. The same applies to the Temple Mount in Jerusalem rebuilt by King Herod.

Carbon-14 dating, for instance, merely dates the organic artefact that has been found. It does not conclusively date the site that surrounds the artefact. Videos and websites such as *Ancient Aliens Debunked* try hard to remove the mystique of these colossal structures that remain, concentrating on their own interpretation of the facts, but the mystery remains. One can believe them, or one can take the views of academics like Arthur Posnansky of the University of La Paz or Oswaldo Rivera, the former Director of Bolivian Archaeology, who believe the site of Tiahuanaco goes back beyond commonly accepted views of when civilisation may have existed there. For one thing, the city of Tiahuanaco underwent massive physical movement, having formerly been a port of gigantic proportions and ended up being marooned twelve miles south of Lake Titicaca, with remnants of its port gates such as stone blocks weighing over 400 tons scattered across the site.

"We were here": Giza, Baalbek and Nasca

The literature on the Giza Pyramids in Egypt, the most famous ancient structures in the world, is too voluminous to discuss in any depth here. However, for readers who wish to obtain an overview on the background to the construction of the pyramids in the area called the Memphite Necropolis (which includes the three Giza pyramids, two large pyramids at Dashour and further pyramids at Zawyat-al-Aryan and Abu Ruwash, and covers about thirty kilometres long by four kilometres wide) an extremely readable book on the subject is *The Orion Mystery* by Robert Bauval and Adrian Gilbert, published in 1994. The three Giza pyramids are believed to have been constructed during the Fourth Dynasty in Egypt (2613 BC to 2494 BC). However, were they conceived, designed or constructed at an earlier date, perhaps from around 10,450 BC when the pattern of Orion's Belt was mirrored on the ground by the Giza Pyramids? Perhaps the designers and constructors of the pyramids came from the areas in the sky where the shafts in the Great Pyramid point to: The Orion Belt and Sirius to the south, and Alpha Draconis and Ursa Minor to the north.

Aside from the ground-breaking research of Messrs Bauval and Gilbert, John Anthony West gives a plausible explanation, based on water weathering evidence, that the pyramids are far older than archaeologists will countenance. As Graham Hancock and Robert Bauval have postulated, perhaps the pyramids were the ultimate expression of our predecessors saying "we were here", by laying out the Pyramids to mimic the precise dispositions of the three stars of Orion's Belt in relation to the course of the Milky Way in 10,450 BC and, thereby, giving future civilisations a precise marker as to when they were there. The same could be deduced from the astronomical layout of the stones at Nabta Playa, one hundred miles west of Aswan in Egypt, which also references the historical position of Orion.

Alternatively, perhaps the Pyramids were a marker for the reasons given by Rand Flem-Ath and Colin Wilson in *The Atlantis Blueprint*. It cannot be chance that the Pyramids were built right at the centre point on the planet, too. The precision of the construction at that time is still a marvel to behold in today's technologically advanced world, with no gaps in stone placement. In an age (conservatively placed at over four thousand years ago) when it has been assumed that no modern engineering or construction techniques were known, an incredible thirty million tons of rock was moved around the western desert near modern Cairo.

Sir Flinders Petrie, in 1880, found that the sides of the Great Pyramid of Khufu were lined up almost exactly with the cardinal points of the compass: north, south, east, and west. The accuracy of the alignment is remarkable when one considers the size of the structure. The orientation towards the cardinal points, the keeping of the base square, and the sloping sides perfect are astounding. The Pyramids can still be considered to be the most magnificent structures on the planet - artistically light years superior to our modern steel and glass structures. Whatever the 'debunkers' say about their engineering and construction techniques, they are so precise that we almost certainly could not replicate those structures today.

The construction of the Kings Chamber, as pointed out by Robert Bauval and Adrian Gilbert, provides an extraordinary example of the precision involved. The granite blocks which make up the walls and ceiling weigh about thirty tons each and are perfectly smooth-faced. The black granite was brought from Aswan in Upper Egypt. No mortar was used in jointing, but, as with the casing-stones on the outside of the pyramid, the blocks were so perfectly cut and fitted that a knife blade will not fit between the joints. Such fine jointing is seen in relation to the limestone blocks on the Great Pyramid, and is an astonishing achievement, but to see it applied with such huge granite blocks is nothing short of miraculous.

As is surmised later, this is the Annunaki "trade mark". Apart from the ease with which they appeared to be able to move over 2.5 million limestone blocks with an average weight of 2.6 tons per stone block, the Great Pyramid appears to encode a deep understanding of Pythagorean geometry, ancient metrology, the mathematical value of *pi* and the geodesic measurements of Earth. Moreover, as Andrew Collins indicates from his research in his book *Gods of Eden*, it seems that ultrasonic drilling techniques may even have been utilised. Where did these skills come from and how did they get lost?

The 'debunkers' who wish to belittle the incredible achievements of the Great Pyramid construction say that it would have been easy to lift the blocks up to the relevant height and place them into the construction. They say that a simple ramp was all that was required, but whilst this might have worked for the lower levels, as the Great Pyramid got higher the ramp would have become longer and steeper and would have to have been of sufficiently solid construction not to collapse under its own weight. At the higher points, the ramp would have needed to be a mile or more long, and would have required as much stone as the structure itself. Suitable cranes are not even available today, so the mystery of its construction persists. Was it constructed with the advanced technological help of extraterrestrials? Perhaps they came from the areas in the sky where the shafts found in the Great Pyramid point to: The Orion Belt and Sirius to the south, and Alpha Draconis and Ursa Minor to the north.

In Issue #38, March/April 2003, *Atlantis Rising* magazine published an article entitled *Archeology and the Law of Gravity*. The article referenced an attempt in the 1970's by a well-funded Japanese engineering team to show that they could match the precision engineering that went into building the Great Pyramid using traditional tools and methods to build a much smaller, sixty-foot scale model of the Great Pyramid. The team constantly ran into embarrassing technical problems. Their problems started at the quarry when they discovered they could not cut the stones from the bedrock and had to call for modern jackhammers to deal with the problem. Then they tried to ferry a block across the river on a primitive barge, but could not control it and had to call for a modern one. When they reached the opposite bank, they discovered that the sledges sank into the sand and they could not move them. They had to call on a bulldozer and a truck. To round off the misery of the Japanese team, when they tried to assemble the pyramid, they found they could not position the stones with any accuracy and were forced to call on the aid of a helicopter to complete the task. As the article, written by Will Hart, points out, the Japanese went home humiliated by the inability to bring the four walls together into an apex. Hart went on to marvel at the inconceivably exact planning that must have gone into building the Great Pyramid in order to bring the 481-foot-high walls to a point. The article concludes that it is absolutely apparent that the ancient Egyptians could not have built the Great Pyramid with the tools and techniques that Egyptologists claim they used.

Perhaps an even greater monument to the past is the structure comprising the solid blocks of limestone which are often known as the Stones of Baalbek in Lebanon. The sheer size of this stonework is unmatched anywhere else in the world, but this wonder is virtually never talked about. The largest stone blocks at Baalbek are around twelve hundred tons, nearly three times as heavy as the heaviest blocks at Giza. Whatever anyone says about the way the blocks were lifted into place at Giza, the blocks at Baalbek could not have been assembled using ramps and pulleys, or other methods suggested by today's archaeologists. They are simply too heavy and, even today, it would stretch the greatest engineering companies in the world to find methods to erect these blocks, and nobody has tried anyway.

On top of the massive stone structure (which was so perfectly worked and spaced that it made placing a structure on top of it simple), the Romans built a temple, which can easily be distinguished from the lower-level base. The two are clearly different constructions. It would require advanced technology to erect the base structure in the way it has been cut and laid, but not the upper structure.

Finally, there are the Nazca lines in the Nazca desert in Southern Peru. Only a constructor with incredible technology and expertise could have flattened and levelled perfectly the tops of mountains in the way it occurred. What were the Nasca Lines created for, and when? They can only be seen in all their glory from the air, which seems impossible until the recent development of aviation.

Perhaps this is a classic case of the Annunaki stating "we were here". Why would they undertake such a task? We don't know. It was obviously something they could do; perhaps, they did it simply because they could, leaving a permanent reminder of an advanced people from before our time.

The Annunaki Trade Mark

As in all megalithic "Annunaki" constructions, each stone fits so close and perfectly with its adjoining stones that the joints are barely visible, and not even a razor blade or piece of hair could fit into the joints. This is the "trade mark" of the constructors, and they used the same technique wherever they constructed, as if to say "this was us" - whether that be the Stones of Baalbek, the Pyramids or Stonehenge (amongst many other masterpieces).

The reality is that an old bygone civilisation used advanced technology to erect these structures, for reasons unknown. But it is a wonder of the long distant past, and we should acknowledge that something special happened, using remarkable technology, to create the structures. From the evidence presented earlier, why should this not have been extraterrestrial technology?

Weightless Technology

There was an interesting observation by the Team Commander of the humans who travelled to Serpo, referenced earlier. He states that forty-five tons of cargo that had been loaded onto the "big ship" on Earth was "offloaded in one big move". This was clearly done quickly and effortlessly. The "ebens" simply moved the whole platform of a mother ship they had landed on. As Len Kasten points out in his book, the entries by the Team Commander in the log settle once and for all the speculation about how the Pyramids of Egypt, Stonehenge, Baalbek, and all the other massive ancient megalithic archaeological sites were built. They simply moved heavy objects utilising some form of weightless technology.

Stonehenge

And what about Stonehenge? According to Zecharia Sitchin's words, Stonehenge was an instrument to measure the passage of time. It was neither a palace nor a burial place, but was in essence a temple-cum-observatory, as the ziggurats (step-pyramids) of Mesopotamia and ancient America were. Modern scientific research methods have confirmed the findings of the renowned Egyptologist Sir Flinders Petrie that Stonehenge dates from about 2000 BC, over four thousand years ago.

Over the last two and a half centuries or more, Stonehenge has been shown to be orientated to the summer solstice, but also has an unmistakable lunar relationship. The main purpose of the design and construction of Stonehenge was to predict eclipses. This made Stonehenge nothing short of an "astronomical computer made of stone".

Leaving aside the immense logistics of how the stones of the original construction of Stonehenge may have been placed, the incredible thing is that the entire Stonehenge circle allows accurate predictions of solar and lunar eclipses. This is revealed in Sitchin's *When Time Began*, in which he referenced the findings of Professor Hawkins of Boston University. The builders of Stonehenge knew in advance "the precise length of the solar year, the Moon's orbital period, and the cycle of 18.61 years."

Nowhere else in the Northern Hemisphere does such a construction occur. It seems that the builders knew the one precise place in the Northern Hemisphere where an open-air "altar" would permit "worship" of the Sun and Moon in their twelve distinct phases over an 18.61 year cycle.

Would this same type of construction occur today? Could it? The answer is almost certainly in the negative.

PART 7

Travelling Aliens:
The Conclusion

The Future of Humanity is not looking good

The introductory chapter of this book highlighted how extensive and unremitting violence and warfare has been since the beginning of human existence. This final chapter returns to that theme. Mankind's history since the dawn of civilisation has essentially been one long bloodbath, with man fighting fellow man over often what are trivial differences. Literature referencing extraterrestrial involvement in this agitation has been reviewed earlier, in particular William Bramley's *The Gods of Eden*. This included the methods that such extraterrestrial 'Controllers' utilised to ensure continuing hardship of humanity.

The reality of the long, unremitting, bloodbath cannot be doubted by anyone who takes the time to read the leading work on mankind's violent tendencies: Colin Wilson's enormous and scholarly work *A Criminal History of Mankind*.

Wilson was a great admirer of H.G.Wells who, in 1898, wrote one of the first widely circulated novels about an alien attack on humans, *War of the Worlds*. Near to the end of his life in 1946, Wells changed his optimistic view about the future of humanity. In the final edition to his *A Short History of the World* in 1945, he added a short 34-page postscript entitled *Mind at the End of Its Tether*. He stated: "Since [1940] a tremendous series of events has forced on the intelligent observer the realisation that the human story has already come to an end and that *homo sapiens*, as he has been pleased to call himself, is in its present form played out". Based on the information contained in the earlier chapter entitled "Where It All Ends", Wells may have reached the right conclusion, but just omitted the extraterrestrial connection.

Colin Wilson commented on Wells' *Mind at the End of Its Tether* by stating that the brutalities of the Nazi period of history may have forcibly changed his view from optimistic to pessimistic. After the brutality of the First World War, it had been assumed by the mass of humanity (including Wells) that no further war of that sort would ever be fought again, and that world government would guarantee peace across the globe evermore. But how wrong they all were. Less than a generation later, the world exploded into even more ferocious violence, and cruelty, in the Second World War: the atrocities of the Japanese, Stalin, Hitler, and the dropping of two atomic bombs. Wells, therefore, stated that "the final end is now closing in on mankind".

Abraham Maslow proposed a basis for human motivation, which he called a "hierarchy of needs", in his 1954 book *Motivation and Personality*. Once one "need" of humanity was satisfied (which Maslow stated followed in the order of food; home; love; and then self-esteem) a person would seek to satisfy the next stage of "need". Colin Wilson saw that this also corresponded roughly with different historical periods of crime. Crimes in society had started with survival as the basis for crime, then had moved to the safeguard of their homes, then sex-related crimes (such as the famous Jack the Ripper crimes) and finally in recent times to crimes related to self-esteem, where people started to feel their dignity, justice and individuality were being compromised. Of course attacking nations rather than individuals was also part of that process. Once the need to be admired had been satisfied, people were free to develop the final stage which Maslow called 'self-actualisation', where we seek out knowledge. Most people, Wilson observes, never reach the final stage.

It seems that, as a breed, members of humanity simply cannot get on with its fellow citizens. Over population is making violence more evident too. We all need personal space, and rarely get enough of it. Desmond Morris even went so far as to state that cities, with their overcrowded conditions, were "human zoos". Morris had studied animals in captivity and, in his 1969 book *The Human Zoo*, he noted that the traits of humans and animals were much the same once animals lived in zoos and humans lived in cities. Both exhibited a variety of perversions which would not normally occur without stressful conditions and, in the worse cases of dominant individuals (or animals), this resulted in violence, cruelty and murder.

In his book *Alien Mind*, published in 2012, George LoBuono took extraterrestrial involvement in humanity's affairs to the point of conclusion reached by H.G. Wells that was mentioned earlier: in other words, that *there is unlikely to be any future for humanity*, at least in its current form. It seems that the negative traits of mankind may have sealed mankind's fate in the eyes of the aliens who have been observing and interacting with humanity for many years. The concepts raised in LoBuono's lengthy book are not easy to follow, but the key concepts raised are nevertheless important.

LoBuono discusses two distinct varieties of alien: the aliens that are ecologically responsible in the universe, and the aliens that are not. According to LoBuono, who appears to have acquired his information through remote viewing and telepathic communication, the inhabitants of Earth are being influenced, abducted and hybridised by irresponsible, power-hungry, acquisitive aliens who do not have humanity's best interests at heart. This is a concept raised in the literature reviewed earlier, such as by Bramley, Sitchin, Jacobs and Turner, amongst others. According to LoBuono, the wiser, older variety of alien who can influence aggressive conduct of younger, acquisitive-minded aliens around the universe, are not prepared to intercede on behalf of humanity because of humanity's reprehensible conduct of its affairs. Hence we are on our own, which does not auger well for us. Indeed, the only conclusion that one can come to is that few, if any, of the current version of *homo sapiens* will survive much longer.

But the most interesting feature raised by LoBuono is how insignificant Earth and its inhabitants are in the grand scheme of things in the universe. The sheer enormity of the universe, the variety of inhabitants of planets within the universe, and the vast distances and light years involved to travel across the universe, is beyond our wildest imagination.

Data released in 2010 indicates that in our galaxy alone (the Milky Way) there are at least 200 billion stars and at least 46 billion earth-sized planets, and astronomers are aware of at least 100 billion galaxies at this time. It would seem reasonable, therefore, to expect vast numbers of different life forms out there. This recent data reinforces the data provided by Richard Preston in his 1987 book *First Light*. Preston's book centred on the development of the Hale Telescope in California, and the search for "quasars". Quasars are considered by astronomers to be the bright light at the edge of the universe, which allows us to measure the boundaries of the universe (currently thought to be between 10-20 billion light years away). These facts and figures make one realise the insignificance of planet Earth in the overall picture, and the absurdity of the trivialities humans worry over.

In fact, it seems that the recent thought of physicists such as Stephen Hawking is that the universe actually doesn't have any boundary, or edge (as Richard Preston discussed as a reason for the search for 'quasars') but, instead, cycles back, or folds into itself via the inward pull of black holes, gravity, and the atomic fusion that brings atoms closer together in all the stars of the universe. Perhaps the universe cannot stretch out indefinitely.

Moreover, some physicists have recently suggested that the universe may have been re-cycled numerous times already, which has created a 'multiverse', in other words, a series of interconnected, or inter-dimensioning universes. LoBuono says that string theorists suggest that a previous "brane", a sheet-like fabric of time existing in a previous universe, could have connected with our "brane" to cause the current universe to form. This may be the basis for the concept of extra dimensions ("branes" as string theorists call dimensions).

For an alien to travel from a distant galaxy to Earth, Colonel Philip Corso said that aliens fold gravity and literally pull two distant points of space-time together for faster-than-light space travel. Hence, aliens living in a galaxy hundreds of millions of light years away can still reach earth quickly with advanced technology. Genetic engineering (not unknown to humanity) is a normal feature of alien societies for numerous reasons, not least for intelligence and hyperspace travel. LoBuono does state, based on his interaction with older aliens, that there are no unknown frontiers within the universe: All galaxies have been surveyed scientifically and those parts not already occupied have been designated for future life. And vast numbers of aliens exist out there. One group LoBuono talks about ('The Verdants' as written about by Philip Krapf, the respected award-winning *Los Angeles Times* journalist) have a population of *500 trillion* on their planets (these aliens apparently inhabit a significant number of planets).

To most humans struggling to survive on Earth, some of the information reported by LoBuono may stretch the boundaries of our minds. But LoBuono states (as other writers reviewed earlier in this book have also stated) that there are aggressive aliens from far away galaxies who are mentally and technologically far superior to the Earth-based human, and who do not have the best interests of humanity at the forefront of their thought processes. These aliens are currently abducting humans and creating human hybrids for their own self-serving reasons, not least the overpopulation of their own planets which is causing them to seek new pastures and resources.

The problem humanity has to face, according to LoBuono, is that the current version of the human is scheduled for discontinuation once the appropriate hybrid, which is currently under preparation by these aggressive aliens, is ready. The breeding program is well underway. The older alien forms are not inhibiting these aggressive aliens because the human is not considered worth saving in its current form.

Humans are currently making planet Earth uninhabitable. We allow poverty, crimes against humanity, and ecological disaster. Additionally, we are self-serving, greedy, aggressive, reckless and thoughtless of others. The wider alien community consider this constitutes a threat to other alien worlds.

The end result is that the wider alien community is of the opinion that humans can be abducted and hybridised with impunity. There is no long term future for our children unless we learn to control the current negative human traits, but it may be too late already. Certainly, even if we had the technology, LoBuono says he understands that there is no other planet we could move to if we trash this planet – humans deriving from planet Earth would be "homeless".

LoBuono says it has been suggested by advanced aliens that the future for some humans may be to move to a more advanced kind of universal consciousness and exist as a higher life form, as the current human will likely cease to continue in its current form as this current life form is too dangerous to the wider alien community, and will become extinct.

Whilst this all may seem somewhat farfetched and worrying, particularly as the majority of the vast global population of earth has yet to come to terms with the idea of extraterrestrial existence, in the end the future of humanity may come down to choosing between the continuation of its current pursuit of material possession of things, power and control, *or* the more laudable aim of equality, transparency and sharing.

Perhaps it is time humanity turned the telescope round and looked outwards, and grasped the implications of our lives in the wider universe, rather than the current narrow-minded inward-looking, self-serving perspective of much of humanity. Humanity's saving grace may be its capacity to evolve through intelligence, but time is short and Earth's leaders rarely exhibit characteristics which could save mankind from extinction.

Addendum

We all want to know why we exist, why we are here, who we really are, where we came from, and whether there are other forms of life out there that may help or hurt us.

What we can understand from the information presented in this book is that we are not alone. Something more sinister is also planned though. This is that someone or something is trying to remove our freedoms, freedom of movement, freedom of thought, our spirit, our free will.

The reality is that we are controlled as a race and have lost our spiritual way. Earth is being trashed, and we don't see it, or don't care, because we are trapped, lacking in awareness, and have little interest in life beyond the small world we live in. It seems that the wider 'world' is irrelevant. Materialism has been made fashionable and desirable by social media and advertising, with the intention that we should never look beyond the material universe. This makes manipulation by the rulers of society easier.

Have we deliberately forgotten the existence of a Supreme Being? We need to find the happiness which has been lost to us as a result of the control of humans and society at large. We appear to have little understanding of, or even desire to understand, any world which is not visible and immediately accessible with our five senses. Worse still, we spend most of our lives trapped by society and unable to see beyond the everyday process of survival. Most people finish school, find a job, get married, have kids, struggle to create a home and pay their debts, which takes all their thoughts and energies (and is what society expects) and then, when life is ending, it is not unusual to look back over time and say: *Was that it?*, *Was that all?*

And then we die unsatisfied, having learnt nothing, and having spent all our lives as slaves to the government, society, and big corporations.

As Colin Wilson has noted as a theme in many of his books, such as *The Outsider*, just occasionally, we escape the futility and meaninglessness of our everyday lives. Only when we are faced with a crisis do we understand the meaning of freedom, and then grasp how easy it would be to live on a far more intense level of vitality and purpose. Yet, almost as soon as the crisis vanishes, we sink back into the 'triviality of everydayness', as Heidegger said, and back into the struggle to find a sense of purpose again. Crisis, or a sense of the forbidden, or excitement such as a new love, are ways in which humanity can leave behind, at least temporarily, the sense of futility with life.

To redress the balance caused by the relentless pressure of our daily lives, to make sense of what we are doing here, there are unseen factors we must understand. We need to change our perspective on life. In this way we can recover control over our lives, which involves recovery of our spiritual understanding.

As mentioned at the start of this book, the Dalai Lama made an important statement about death, which is ever present as a theme in this book. He stated, "Naturally most of us would like to die a peaceful death, but it is also clear that we cannot hope to die peacefully if our lives have been full of violence, or if our minds have mostly been agitated by emotions like anger, attachment, fear. So if we wish to die well, we must learn to live well: Hoping for a peaceful death, we must cultivate peace in our mind, and in our way of life".

The Pyramid Texts

Understanding death is important to understanding life. The two are linked in a continuous cycle. Ever since the beginning of modern civilisation, death has been revered more than life, because the individual that was dying was considered to be returning to his true world, the spirit world. This can be seen from the oldest religious writings yet discovered in the world, known as the *Pyramid Texts*, which were extensively reviewed by Robert Bauval and Adrian Gilbert in their 1994 book entitled *The Orion Mystery*. As the authors point out, given their extraordinary antiquity, it is strange that the *Pyramid Texts* are almost unheard of by the general public. They are considered, even conservatively, to be over 5,000 years old, and were written at least three-and-a-half thousand years before the Christian gospels were written and over two millennia before the Old Testament. The texts appear to be related to, and used in, a period significantly prior to the time they were written. They were discovered in the pyramid of Unas, the last king of the Egyptian Fifth Dynasty, and appear to refer to a religion and rituals in existence during the Fourth Dynasty – the period during which the massive and wondrous pyramids of Giza and Dashour were constructed. One of the key features of the texts was the afterlife destiny of the Egyptian kings.

Bauval and Gilbert point out in *The Orion Mystery* that the Ancient Egyptians were religious people and believed strongly in an afterlife. To help the dead reach the afterworld, it was deemed important to preserve the body of the deceased as far as possible and to provide the departed with the "means and accessories for the arduous journey into eternity". By the Second Dynasty, almost five thousand years ago, dead kings were being buried in homes that were more elaborate than those they had lived in. Bauval and Gilbert quote from a book considered to be the definitive study on the Egyptian Pyramids, Dr Edwards' *The Pyramids of Egypt*, where Edwards advances the idea that: "In a land where stone of excellent quality could be obtained in abundance, it may seem strange that the rulers and governing classes should have been content to spend their lives in buildings of inferior quality to their tombs. The Ancient Egyptian, however, took a different view: his house or palace was built to last for only a limited number of years....but his tomb, which he called his 'castle of eternity', was designed to last forever."

By the Third Dynasty, this concept became more elaborate with the creation of the step-pyramid, the largest remaining being King Zoser's at Saqqara. Its ziggurat-style structure seems to have symbolised a ladder whose six steps leading up to a seventh platform may have corresponded to the stages of ascent through which the soul must pass after death. The step-pyramids were designed to be seen from afar and their external look was as important as their internal burial chambers.

The design has been repeated all over the world, and shows a ladder from which the afterlife can be reached. The burial places were becoming more and more elaborate until by the time of the Fourth Dynasty, we saw their culmination in the most beautiful architecture ever constructed, which aligned with the stars.

Dimensions Beyond Life

In her spiritual masterpiece *Initiation*, Elizabeth Haich has said that we occupy a physical body but the immortal body is the spiritual body or being that we also occupy. This could be said to be the real 'you'. Most religions agree that the spiritual body or being is an entity that is aware, creative and has a personality. It is not, however, composed of matter or of any other component of the physical world. It is an immortal unit of awareness that cannot perish but can, and does, download itself into physical matter for periods of time. It follows from this analysis that death is little more than a spiritual abandonment of the body it occupied for a period of time.

As a result of his 1981 survey, George Gallup Jr stated in his book *Adventures in Immortality*: "A growing number of researchers have been gathering and evaluating the accounts of those who have had strange near-death encounters, and the preliminary results have been highly suggestive of some sort of encounter with an extra-dimensional realm of reality. Our own extensive survey is the latest in these studies and is also uncovering some trends that point toward a super parallel universe of some sort".

In the Greek, Egyptian and other ancient eras, it is reported that people in secret societies were given 'controlled' NDE's, chemically induced psychedelic experiences on par with the accidental nature of NDE's. These experiences were kept secret and to this day are referred to as "mysteries", such as the Greek Eleusinian Mysteries. These are explorations of a reality that exists beyond the threshold of what we see and experience in our everyday reality. The 1990 film *Flatliners* is a good introduction to the concept of chemically induced NDE's, which assist growth of the spirit.

Michael Talbot, in *The Holographic Universe*, states that NDE's often describe going through a "passageway to the land of the dead". He argues that the similarities between near-death experience and the *Egyptian Book of the Dead*, an ancient Egyptian funerary text from nearly 4,000 years ago, which is sometimes translated as "The Book of Emerging into the Light", are more than coincidence. That book documented the journey of the soul through the underworld and into the afterworld, or afterlife. To the Egyptians of that era there was nothing more important to them in life than accomplishing this journey in death.

Thought and Consciousness

On the basis of all the literature reviewed at the start of this book in connection with the creation of the human being, it is not a great leap of faith to hold a belief in the guiding hand of a Supreme Being.

Indeed, the highly respected anthropologist and natural science writer, the late Loren Eiseley, who was called a magus, spiritual master or shaman by Dr Richard Wentz, wrote, "We have every reason to believe that, without prejudice to the forces that must have shared in the training of the human brain, a stubborn and long drawn out battle for existence between several human groups could never have produced such high mental faculties as we find among all peoples on the Earth. Something else, some other educational factor, must have escaped the attention of the evolutionary theoreticians".

The key component of the guiding hand in creation may very well be *consciousness.*

In his book *The Secret History of the World*, Jonathan Black discusses the secret societies that have passed down to their initiates the true meaning of life. Black summarises much of what key esoteric figures across history passed down to initiates. He says that the real secret is that *mind preceded matter.* Science says that before the big bang there was nothing, no objects in space, no time. It was *thought* that generated physical events, which all started from the "point of singularity". And as the secret societies understand all too well, *thought* continues to be the key to creation and existence.

In fact it has been said that all we really are is consciousness, and that this is what is released at death to allow us to experience the limitless freedom that awaits us, which is restricted from us in this physical world. Dutch cardiologist Pim Van Lommel produced a massive study of near-death experiences that supported the whole concept of life after death. Through this work, Lommel stated he now saw that "everything stems from consciousness." He went on that he now understood that "you create your own reality based on the consciousness you have and the intention from which you live".

Philip Coppens, one of the well-known presenters on the *Ancient Aliens* television series, and who is mentioned later in the chapter entitled "The Monuments" in relation to his book *The Lost Civilisation Enigma*, noted that the presence of consciousness, separate from the body, is what quantum physics has been trying to make humanity aware of for several decades.

Any discussion of "thought" takes us to the heart of the unseen world. Many still do not realize that the solid world we see has little relevance in shaping our world, because it is already here. The world that actually shapes the world we exist in today is invisible. The world we can see is fixed and solid, and this makes us feel comfortable, but the spiritual world is invisible, so we find this more difficult to relate to. Thought is not something we can 'touch', but the way we think creates the world we live in. This was a key point made by Jonathan Black in his classic work.

The idea set out below that 'thoughts create your world', is not just a new age-style concept, but scientific reality as a result of the research conducted by people such as David Wilcock in *The Source Field Investigations*.

When Rhonda Byrne's book *The Secret* came out in 2006, it was sensational, selling millions of copies. Rhonda Byrne went on to be named in Time Magazine's list of the top one hundred people who shape the world. The essence of *The Secret* is that whatever our dominant thoughts are will dictate the world we create around us. This reality has been passed down the ages of secret societies and intellectuals - ranging from Plato, Shakespeare and Newton to Einstein. The book by Rhonda Byrne makes this 'secret' available to everyone.

Parallel Universes

Some people, such as David Wilcock in his book *The Source Field Investigations*, have written of a living energy field from which the entire Universe is built, and the concept that we are lovingly guided by a hidden intelligence. Wilcock explains that a 'source field' may well be operating through our DNA. This he demonstrated from a discovery made by the scientist Dr Peter Gariev, where it seems that there is an energy field that is paired up with every single DNA molecule, as if the DNA molecule has an energetic 'duplicate'. By a simple extension of Dr Gariev's study and experiment, it is reasonable to project that our entire body must have an energetic duplicate. This is essentially an information field that tells our cells what to do and where to do it.

David Wilcock explains the concept of an energetic duplicate, or information field, or better still, the 'parallel universe' we are individually linked to, from his science sources: "While the physical body is in the physical universe, we communicate with ourselves from the parallel universe, and guide the brain to do work and live our lives. Once we die, we continue to live, except we withdraw from the physical universe since the body is not usable. We continue to live in the parallel universes." He continues: "Some scientists now believe that after death, the electromagnetic spatial source of energy, or the soul, just moves through a tunnel of decelerating time to end up at the white light – which represents the entry to the parallel universe. According to researchers in this field, an advanced civilisation can achieve immortality by being able to access the physical and parallel universes anytime. Most probably that is what happens when we are born and when we die. But the technology of accelerating time or decelerating time will allow us within the physical universe to access and move in and out of the parallel universe."

It seems that our individual support teams which assist us during our physical incarnation operate at a higher-dimensional level that defies all known laws of physics that hold true in the physical universe. Wilcock continues from his sources: "Time ceases to exist, making living in that dimension very different from the physical universe, unlike in the physical world, in the parallel universe we can walk from one time to the next time." Of course in the physical third dimensional world we exist in we cannot move from one time to another, we have to endure the linear time gap first before the next 'time' or significant 'event' can arrive. Wilcock goes on to quote from his source in India scientifically how time travel works through the creation of negative mass: "Once negative mass is created, all the puzzles of time travel, bending space and time, movement in and out of a parallel universe – all can be instantaneously solved." This may be how higher dimensional beings travel, such as the UFO's which seem to appear and disappear so easily: "As you enter a black hole, if you can accelerate the process of making your mass negative you can easily pass through...the on-board computers [on a spaceship] are able to control the mass factor from positive to negative and so on just like airplanes balance weight during takeoff, flight and landing. Once mass of the entity can be manipulated, the travel through wormholes will be easy. This will not only enable us to travel to different time dimensions but also into parallel universes and beyond."

Changing everyday thinking in this material, spiritually-lacking world is a hard task, and the concepts raised above are not for everyone. However, there are attempts being made to bring 'way-out' concepts such as parallel universes into more mainstream use or thinking, which can be seen in mainstream films like the *The Golden Compass* and *The Hitchhiker's Guide to the Galaxy* which are about the existence of parallel universes, and mass audience films like *Existenz* and *The Matrix*. We must believe their existence, the existence of the spiritual world, the world beyond the physical world, even when 'friends' and colleagues laugh, mock or ridicule in their attempts to limit such thought processes.

Finally, David Wilcock brings us back to the way the parallel universe works for us at the physical level where we exist now, quoting from his sources: "Two-thirds of our brain is not under our control. It is guided by the entities in the parallel universe. We communicate using mind-generated signals commonly known as telepathy with many other entities, and even with ourselves in higher spatial dimensions of the parallel universe...we do traverse the parallel universes with our mind all the time without physically leaving the physical universe. That part of the brain is not even under our control. It therefore seems that we are connected genetically to some advanced life-forms that can traverse from the physical to the parallel universe all the time."

The New Earth

It would not be hard to conclude from the reviews within this book that the human race and planet Earth itself are struggling for survival at this time.

A lighter but somewhat sensational scenario concerning this situation is suggested by Dolores Cannon within her series of four books published between 2001 and 2012 entitled *The Convoluted Universe*. This is that a "new earth" is on the way.

This may be hard to imagine for many, but as Dolores Cannon is a well-known writer and lecturer in the wider field of extraterrestrial research, a review of this aspect of her voluminous writings is appropriate as a counterweight to the negative conclusion from the reviews in this book.

She promotes a fascinating and unique viewpoint of the near future. Although Cannon writes about the near future, Cannon's view of the future actually emanates from *regression* hypnosis sessions.

By way of explanation, Dolores Cannon is a regressive hypnotherapist and psychic researcher who records 'lost' knowledge. Over a period of many years, Cannon developed a unique hypnosis technique which takes the subject into the somnambulist trance state (the deepest possible trance state) where she believes all the answers can be found. This is because at this deep level of trance, interference of the "conscious mind" is completely removed. This allows what she calls "the subconscious mind" to speak out clearly. A more usual term that other researchers and practitioners have used for this layer of trance state rather than "subconscious" include "soul consciousness" and "supraconsciousness". However, for the purpose of this chapter, the term "subconscious" will be used, as this is the term that Cannon uses.

Cannon states that she takes a subject to a past life that holds the answers to the problems of that subject in this life, and sometimes the answers received relating to a subject have a wider significance than simply curing the subject of the immediate problem. The answers have sometimes given warnings or statements on the wider issues of humanity and its future survival.

Cannon is of the view that the "subconscious" is a record keeper, recording everything that has ever happened in a person's life, including existences in the spirit state. That information is stored for many reasons, but if we were able to listen to the deeper layers of our minds more directly, without the interference of our rational mind, then we would hear lots of messages and warnings that we often miss because we are too focused on our everyday lives to notice even obvious things such as backaches or headaches. We just ignore them when, in fact, the deeper layer is trying to talk to us.

During the hypnotherapy sessions conducted by Cannon, the subjects often talk through their "subconscious", and when this happens the "subconscious" will talk about the subject in the third person (he, she). The "subconscious" is detached from the problems and talks as an objective observer. This causes many interesting things to be revealed, as Cannon shows us in the transcripts set out in the four volumes of *The Convoluted Universe*.

It appears that people accumulate great knowledge from their past lives. Sometimes the information is remarkable. It is assumed that all past life knowledge is 'lost', but through the process of the deep trance hypnotherapy, Cannon discovered over many years of practice that this is not true. Anything that has ever been learned in another life, any talent or skill, is never lost. It is stored in what she refers to as the "subconscious mind", and can be revived and brought forward to be used in the present lifetime if it is appropriate. Healing knowledge, for instance, is often being brought forth again at this time, as it will be needed in the time we are heading into: a time of change.

The coming changes

In the sessions reproduced in Volumes 2 and 3 of *The Convoluted Universe*, certain subjects spoke through their "subconscious" about the *coming* changes on Earth, even though the sessions were *regressive*. This was a reason why some subjects had returned to Earth, or were visiting Earth, at this time. It seems that many beings throughout the universe are watching what is occurring now on Earth and are of the view that something unique is happening. The transcripts indicate that it is the first time any planet or civilisation has gone through the events that are happening now, when an entire planet will reach the level where it will increase its frequency and vibration to allow it to shift en masse into another dimension.

In *The Convoluted Universe*, it is revealed that there is one problem with the coming dimensional shift: Not everyone currently on planet Earth will make the shift. Those that have not achieved the required vibrational frequency will remain on the old Earth. These people will not experience the peace and love that will radiate from the new Earth. They will remain on the old Earth with its wars, killing, greed, hatred, and all the other negative traits that currently make life so painful for many of the human race. Vibrational change takes time, it cannot be done in a second, and there is little time left now before the shift is complete. Those who wish to be part of the new Earth must act now. The transcripts from the hypnotherapy sessions indicate that society will break down and the existing structures crumble. Many will not realise they have made the shift, though they may sense it.

Love and Conduct

The key to making the transition, it would seem, is that we need to operate from a standpoint of *love*. That should be our guiding motivation. Love is about *conduct* - how we conduct ourselves. True love is shown by the way we treat others. Do we think about others as we go about our daily lives, or are we primarily self-serving? One's vibrational frequency cannot increase if we are motivated by greed, jealousy, envy, hatred, hubris, desire for control and reduction of other's free will, and the many other negative human traits that most humans still exhibit. As Dolores Cannon writes in the New Earth section of the transcripts, "We can change the world, one person at a time. Love is the answer, it is that simple". If we are working for the Earth, we will be provided for. The driving force now must be love and service, not greed.

Diet

Cannon indicates that throughout her work, her subjects are regularly told that they must change their diets in order to make the adjustment into the new world. Our bodies must become lighter, and this means the elimination of heavy foods. During the hypnotherapy sessions, the subjects of Cannon are repeatedly warned to stop eating meats, mainly because of the additives and chemicals that are being fed to the animals. These are being transferred to our bodies and remain deposited in the organs for a long time, and are difficult to eliminate. We are especially cautioned against eating animal protein and fried foods, which act as an irritant to the body. One response from the subconscious of a subject stated: "These act as aggravators to your system after many years of misuse. We do not mean to be judgmental, but the body is built for a certain type of vehicular traffic. The body cannot ascend in frequency to higher dimensional realms if the density and the toxins are polluting the environment of the human body". Water is also reported to be extremely important because it makes up seventy percent of our bodies.

Why the shift?

The transcripts indicate that the shift is occurring because mankind has reached the point where the mind-body-spirit balance and appropriate human conduct cannot be restored before the DNA that our extraterrestrial (ET) founders gave humans is irrevocably corrupted. The shift will realign the planet and cleanse and clear the genetic structure of everything and everyone on the planet.

Cannon explains that the various universes are so interwoven and interconnected that if the rotation or trajectory of one is disturbed, it affects all the others. In the extreme case, this could cause all the universes to collapse on themselves and disintegrate. This is one of the reasons for the monitoring of planet Earth by the ET's: to detect any problems caused by our negative influences and alert the other galaxies and universes so countermeasures can be initiated. She explained that they have to know what the Earth is up to so that the other universes, galaxies and dimensions can protect themselves and survive.

Cannon reports that the time is now approaching when a perfect human can be realised and that this is the main reason for the sampling and testing that the ET's have been doing, which has been interpreted as negative. The ET's are concerned about the effects of pollutants in the air and chemical contamination of the food on the human body and, thus, are taking steps to alter these effects. The body is becoming damaged, and so the genetic structure is being altered to make it more perfect. This will allow future humans to be in perfect harmony: not just with themselves, but with the rest of the cosmos. However, the old Earth appears to have reached the point where it is incapable of being helped or changed.

There will be a transition as some attempt to help others make the shift. Many souls are here to help and witness the shift. Belief systems and controls will break down. As Cannon quotes from the sessions: "The ones who are prepared to see these changes and not crumble in fear will be the pillars on which others will lean when nothing makes sense".

Hold on for the ride of your life. We may be witnessing and participating in a unique change on Earth if the transcripts and writings of Dolores Cannon are correct.

The Unseen World

The unseen world, which is another dimension of reality, sometimes called the spirit world, vibrates at a speed too fast for us to see. We can only see 'luminous matter', which reflects electromagnetic light, but 95% or more of the mass of the universe is known as 'dark matter' which doesn't reflect light and so cannot be seen. Indeed, the electromagnetic spectrum is only 0.005 percent of the estimated mass of the universe and human sight can only see a minute fraction even of these electromagnetic frequencies, as we can only see 'visible light'.

With this knowledge of our visible limits, as David Icke points out in *Infinite Love is the Only Truth – Everything Else is Illusion*, it makes a mockery of the scientific view that there are no other forms of life in the universe. How would 'almost-blind' humans really know anything about the rest of the universe? Other alien forms may well operate in realms unseen by the human eye. If such 'alien forms' are ever sighted by humans, it is simply because they entered the frequency range of humans. And when they are no longer visible, it is because they have left our visible frequency range.

All this information leaves us with one final thought. With the now-known marvel and complexity relating to the genetic make-up of the human body, surely there must be universal intelligence, the guiding hand of a Supreme Being, in the process?

Sovereignty of the Soul

Throughout this book there are references to the need to retain our individual sovereignty and free will. In order to assist this process, we need to understand that there are important aspects of our existence that are unseen, but should not be discounted just because they are outside our immediate vision.

Much of what affects us as we live our lives on Planet Earth is, as mentioned above, *unseen*: thoughts, energy, air, viruses, gases, and so on. But above all, our souls are *unseen*. The soul is the intended inhabitant of the physical human body, which is sometimes called a 'body-mind unit'. The body is simply the vehicle used by the soul so that it can experience another environment other than its home base (though some of the soul's energy remains at its home base), in this case the three dimensional environment called Planet Earth. There are many other dimensions, which offer different experiences as the soul develops, but what the soul is experiencing through the body it inhabits as you read this is '3D Planet Earth'. Philosophers such as Rudolph Steiner and researchers such as Linda Moulton Howe have expressed the view that the soul of a person passes in and out of 'containers' (the human body) to help the soul grow. Knowledge is accumulated through different 'containers' the person has utilised in each life cycle.

When the soul incarnates into a physical form which it will inhabit during its time on Earth, it is supported during its occupation of the body by its support team, who reside in the soul's home base and have a communication channel open at all times with the soul during its incarnation on planet Earth. This base is at another level of existence, which we can call 'the spirit world'. This level of existence, our 'parallel universe', which was discussed earlier when referencing David Wilcock's work, gives it 'substance' from a scientific perspective. But even without any backing from 'science', spiritual healers understand this concept intuitively.

The reality of recent findings in the world of quantum physics should not be overlooked in relation to the existence of a 'parallel universe' where it has been discovered that there is instantaneous communication faster than the speed of light between different measured points, and the interaction does not diminish with distance. Einstein and others have measured particles at one place which instantaneously determine the properties (position, momentum, and spin) of a particle detected at another place. This has relevance (from a scientific perspective) to the existence of our support teams. Dion Fortune referenced this in *Psychic Self Defence*, where she notes the findings of quantum physics which confirm the proposition that our world is supported by an unmediated invisible reality (in other words, an invisible reality which doesn't have anyone or anything intervening or acting as an intermediary). Deepak Chopra, the physician-turned-spiritual-writer, in a chapter entitled *Matter, Mind and Spirit* from his 2003 book *Synchrodestiny*, perfectly describes what he calls the "virtual domain", and it is worth reading his book for that chapter alone. He says, "The virtual domain is not a figment of the imagination, the result of some human longing for a universal force greater than ourselves. Although philosophers have been discussing and debating the existence of "spirit" for thousands of years, it wasn't until the twentieth century that science could offer proof of the existence of nonlocal intelligence." He then proceeds to brilliantly summarize the scientific evidence.

Finally, how we conduct ourselves in life across the spectrum of activities we undertake is vitally important. We generally don't realise that it doesn't begin and end with this life; there was a past and there will be a future, because the soul is immortal - unlike the body. Indeed, as Dr Whitton pointed out in his book *Life Between Life*, the smoothest transition from the incarnate to the disincarnate state is accomplished by those individuals who have spent their lives molding an outer character in accordance with the soul's highest impulses. They rejoice over the body's disintegration and are exhilarated at the prospect of being free from encasement. This is a clear countermeasure to the circumstances described in the earlier chapter "Global Control of Humanity".

Once it is understood that we are souls, living inside a body-mind unit and that, although the human body will die, our soul is immortal, then we can start the process of properly living, learning and experiencing in our physical form, which is what we came here to do. Material possessions are solely a teaching tool; they are irrelevant otherwise. The only thing we take back with us to the spirit world is our knowledge and learning - nothing more. And that is enough.

Acknowledgment

Thanks to Angel-Light Love, who is based in Dallas, Texas, a true healer and teacher, for her help in reviewing my many drafts of the book as it developed, and for her support over the years. She is someone who really understands the meaning of the words "service-to-others". Her Angel-Light Beamer blogs can be accessed at http://angel-light-love-healing.blogspot.com

Bibliography

Bauval, Robert and Gilbert, Adrian – "The Orion Mystery"
Begich, Nick and Manning, Jeane – "Angels Don't Play This HAARP"
Begich, Nick – "Controlling the Human Mind"
Behe, Michael – "Darwin's Black Box"
Behe, Michael – "The Edge of Evolution"
Black, Jonathan – "A Secret History of the World"
Boulay, RA – "Flying Serpents and Dragons"
Bramley, William – "The Gods of Eden"
Brown, Courtney – "Cosmic Voyage"
Byrne, Rhonda – "The Secret"
Calleman, Carl Johan – "The Mayan Calendar and the Transformation of Consciousness"
Cannon, Dolores – "The Convoluted Universe" (four volumes)
Cathie, Bruce – "Mathematics of the World Grid"
Chopra, Deepak – "SynchroDestiny"
Collins, Andrew – "Gods of Eden"
Condon, Richard – "The Manchurian Candidate"
Cooper, Gordon – "Leap of Faith"
Cooper, William – "Behold a Pale Horse"
Coppens, Philip – "The Lost Civilization Enigma"
Cori, Patricia – "No More Secrets, No More Lies"
Corso, Colonel Philip – "The Day After Roswell"
Doherty, Robert – "Area 51" (the mass paperback series)
Dolan, Richard & Zabel, Brice – "A.D. After Disclosure"
Duncan, Robert – "Project: Soul Catcher"
Emoto, Dr Masaru – "The Hidden Messages in Water"
Farrell, Joseph P – "Covert Wars and the Clash of Civilisations"

Flem-Ath, Rand, and Wilson, Colin – "The Atlantis Blueprint"
Fort, Charles – "The Book of the Damned"
Fortune, Dion – "Psychic Self Defence"
Goldberg, Dr Bruce – "Time Travellers from our Future"
Good, Timothy – "Need to Know"
Good, Timothy – "Alien Base"
Good, Timothy – "Earth: An Alien Enterprise"
Haich, Elizabeth – "Initiation"
Hancock, Graham – "Fingerprints of the Gods"
Hancock, Graham – "Supernatural"
Harris, Paola Leopizzi – "UFOs: How Does One Speak to a Ball of Light?"
Hartmann, Thom – "The Last Hours of Ancient Sunlight"
Hitchens, Christopher – "God is not Great"
Horsley, Air Marshall Sir Peter – "Sounds From Another Room"
Howe, Linda Moulton – "Glimpses of Other Realities: Vol 2 – High Strangeness"
Icke, David – "Infinite Love is the Only Truth- Everything Else is Illusion"
Icke, David – "A Guide to the Global Conspiracy"
Icke, David – "Children of the Matrix"
Jacobs, David M. – "The Threat: Revealing the Secret Alien Agenda"
Kasten, Len – "Secret Journey to Planet Serpo"
Kasten, Len – " The Secret History of Extraterrestrials"
Knight, Christopher, and Butler, Alan – "Who Built the Moon?"
Koire, Rosa – "Behind the Green Mask – UN Agenda 21"
LoBuono, George – "Alien Mind"
O'Brien, Cathy – "Trance: Formation of America"
Orwell, George – "1984"
Perkins, John – "Confessions of an Economic Hitman"
Pope, Nick – "Open Skies, Closed Minds"
Pope, Nick – "The Uninvited"
Preston, Richard – "First Light"
Preston, Richard – "The Hot Zone"
Prouty, L. Fletcher – "The Secret Team: The CIA and its Allies in Control of the United States and the World".

Pye, Lloyd – "Everything You Know is Wrong"

Redfern, Nick – "The NASA Conspiracies"

Salla, Michael – "Exposing US Government Policies on Extraterrestrial Life"

Sauder, Richard – "Underwater and Underground Bases"

Schulze-Makuch, Dirk & Darling, David – "We Are Not Alone"

Sherman, Dan – "Above Black"

Sitchin, Zecharia – "The Twelfth Planet"

Sitchin, Zecharia – "When Time Began"

Steiger, Brad and Sherry Hansen – "Real Aliens, Space Beings, and Creatures from Other Worlds"

Stonehill, Paul and Mantle, Philip – "Russia's Roswell Incident"

Talbot, Michael – "The Holographic Universe"

Thomas, Robert Steven – "Intelligent Intervention"

Turner, Dr Karla – "Taken – Inside the Alien-Human Abduction Agenda"

Walden, James – "The Ultimate Alien Agenda"

Warren, Larry and Robbins, Peter – "Left at East Gate"

Whitton, Joel L, MD and Fisher, Joe – "Life Between Life"

Wilcock, David – "The Source Field Investigations"

Wilson, Colin – "A Criminal History of Mankind"

Wilson, Colin – "The Outsider"

Wood, Dr Robert M (with Nick Redfern) – "Alien Viruses"

Lightning Source UK Ltd.
Milton Keynes UK
UKOW05f0156200517
301575UK00008B/41/P